The Open University

Technology Foundation Course Units 7-8

The week number during which this unit should be studied is not necessarily the same as the unit number. Please consult your wall-chart study guide to the Technology Foundation Course to find the place of this unit in the course.

ELECTRICITY AND MAGNETISM

Prepared by the Course Team

D1341154

THE OPEN UNIVERSITY PRESS

The Technology Foundation Course Team

G. S. Holister (*Chairman and General Editor*)
K. Attenborough (*Engineering Mechanics*)
R. J. Beishon (*Systems*)
D. A. Blackburn (*Materials Science*)
J. K. Cannell (*Engineering Mechanics*)
A. Clow (*BBC*)
G. P. Copp (*Assistant Editor*)
D. G. Crabbe (*Course Assistant*)
C. L. Crickmay (*Design*)
N. G. Cross (*Design*)
E. S. L. Goldwyn (*BBC*)
J. G. Hargrave (*Electronics*)
R. D. Harrison (*Educational Technology*)
M. J. L. Hussey (*Engineering Mechanics*)
A. B. Jolly (*BBC*)
J. C. Jones (*Design*)
L. M. Jones (*Systems*)
R. D. R. Kyd (*Editor*)
J. McCloy (*BBC*)
R. McCormick (*Educational Technology*)
D. Nelson (*BBC*)
C. W. A. Newey (*Materials Science*)
S. Nicholson (*Design*)
G. Peters (*Systems*)
A. Porteous (*Engineering Mechanics*)
C. Robinson (*BBC*)
R. Roy (*Design*)
J. J. Sparkes (*Electronics*)
R. Thomas (*Economics*)
G. H. Weaver (*Materials Science*)
P. I. Zorkoczy (*Electronics*)
and the late Professor R. K. Ham (*Materials Science*)

The Open University Press
Walton Hall Milton Keynes

First published 1972   Reprinted 1974 (with corrections)

Copyright © 1972 The Open University

Designed by the Media Development Group of the Open University

Printed in Great Britain by
Martin Cadbury, a specialized division of Santype International
Worcester and London

SBN 335 02505 6

Open University courses provide a method of study for independent learners through an integrated teaching system, including text material, radio and television programmes and short residential courses. This text is one of a series that makes up the correspondence element of the Foundation Course.

The Open University's courses represent a new system of university-level education. Much of the teaching material is still in a developmental stage. Courses and course materials are, therefore, kept continually under revision. It is intended to issue regular up-dating notes as and when the need arises, and new editions will be brought out when necessary.

For general availability of supporting material referred to in this book, please write to the Director of Marketing, The Open University, P.O. Box 81, Milton Keynes, MK7 6AT.

Further information on Open University courses may be obtained from The Admissions Office, The Open University, P.O. Box 48, Milton Keynes, MK7 6AB.

1.2

Contents

Aims and Objectives

The aim of these two units is to introduce you to the principal phenomena in the field of electricity and magnetism and some of their technological applications. The treatment is almost entirely non-mathematical, but sufficiently quantitative to bring out the nature and magnitude of the most important units used in electrical measurement.

After working through the units you should be able to:

(1) State the fundamental law of attraction (or repulsion) between electric charges and calculate the force between two given charges a given distance apart.

(2) Explain electric current and potential difference in terms of the flow of electrons, and account qualitatively for the difference between metals and insulators.

(3) Explain qualitatively how an electric battery operates.

(4) Describe the process of electrolysis and how it is used industrially.

(5) Name the units of electric charge, electric current, electric potential, electrical resistance, electric power and magnetic field (magnetic induction); state the relationships between these.

(6) State Ohm's law and explain the conditions under which it applies.

(7) Draw and interpret simple circuit diagrams to represent networks of resistors and batteries; use Ohm's law to calculate the equivalent resistance of such networks; calculate the current, potential difference and power dissipation in each branch and the overall power dissipation in the network.

(8) Explain what is meant by a magnetic field and how an electric current can give rise to a magnetic field.

(9) State the law of force on a current-carrying conductor placed in a magnetic field.

(10) Explain how an ammeter and a d.c. electric motor operate.

(11) Explain what is meant by electromagnetic induction and state the formulae giving the magnitude of the induced potential difference.

(12) Describe a d.c. dynamo, an alternator and a transformer and explain how each works.

(13) Discuss the advantages of a.c. over d.c. for a public electrical power supply system.

What you have to do

Units 7 and 8 have been written as a single entity covering two weeks work. The written units are almost entirely descriptive. There are 15 simple numerical exercises at the end which will enable you to test your understanding and retention of the work. Probably the most important aspect of the fortnight's work (and the part you will find most interesting) is the experiments you will carry out with your home experimental kit: you will need some graph paper to display your results. The experiments will probably take you between 2 and 6 hours, depending on your previous experience. We suggest you intersperse experimenting with reading the text. Some of the experiments have more advanced tailpieces which you may omit if you are not used to manipulating electrical equipment.

Section 1

Introduction

I have no idea what you expect to find in the units on electricity and magnetism, but if you hope at the end to be able to answer questions like: 'What is electricity?' or 'What is magnetism?' I must warn you now of some coming disappointments. To questions of this form, neither science nor technology can make any useful response. There are of course many other questions to which we can give specific and definite replies—questions like 'How can I recognize electricity?' or 'How can I use magnetism?' It is to finding responses to this sort of question that the present units are addressed.

Electrical phenomena, and their use, are an essential feature of our modern way of life. We use electrical devices in home, office and factory; we use them often and we use them in a wide variety of ways. So interwoven with our daily activity, so much a part of our life, is this dependence on electricity that we may deduce quite a lot about basic electrical interactions merely by referring to things we know well, by thinking of the world which is our common experience. Our studies in electricity will begin from this point of view.

Normal domestic activities give some indication of the range of phenomena encompassed by electricity. Merely by the operation of a switch we can illuminate a room or warm it. We can cook, clean, mix and freeze; we can even cut grass. Although often very different in size, in character or in function we can summarize these devices very easily if we think in terms of output, for all in their varied ways supply us with heat, light, sound and mechanical motion. They are all converter devices, and to get the same outputs from non-electrical devices we should expect to provide coal, gas, oil or even our own bodily effort.

The common ground of these examples then, is that they show the electrical supply system of the country as being primarily an energy supply system, and our domestic appliances as localized units which regulate and control this supply, while turning it into the specific physical form which we require. The wires in your house, under the street or strung overhead on pylons are therefore functionally equivalent to fleets of oil tankers, trainloads of coal, or pipes carrying domestic gas: the household appliances are equivalent to the pistons and pumps, flywheels, shafts and gears which allow us to direct other forms of energy to their point of use. Among all this, the features which contain the uniqueness of the electrical supply are just its speed, its controllability and the simplicity of the coupling between supply and conversion device.

These domestic considerations have of course one great limitation: they are unidirectional. They allow us to discuss only what we can take from the electrical supply and must ignore what goes into it. And something of course must go in: in technology, as elsewhere, you never get something for nothing. In Unit 6, Mechanics and Unit 0, Modelling I you have already met the idea which we need to describe this, and you will meet it again in Unit 20, Energy conversion. Energy is conserved: if our domestic appliances allow us to extract power from the electrical supply, then elsewhere it is being put in.

We can pursue this thought backwards to our power stations. From appliance to fuse box; from fuse box to street; from street to district sub-station; from sub-station to grid; from grid to power station, a continuing connection is maintained and at some point energy must be put in. If we lived in boxes and never looked outside we might be persuaded to believe that electricity is a commodity like energy which is somehow shovelled into the lines for us to extract at one terminal and put back in degraded form at the other. But we know very well that electricity is not like this. Our national fuels are coal, gas, uranium and oil so our power stations are just huge converters which strip the energy from these fuels and turn it to electrical form.

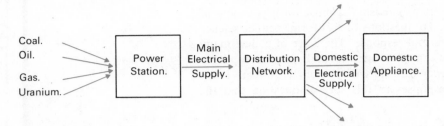

Figure 1

Now nothing in these preliminary thoughts gives any detail about electrical interactions. They merely tell us that they conform to a model like that of Figure 1. This comprises just three elements. The first of these—the power station—accepts large quantities of a fuel like coal and puts something we choose to call electricity into the second element. The second element—the wires of the electrical supply—accepts the output of the first and makes it available at a large number of output points in the same form as it was taken in. The third element—the domestic appliance—accepts the output of the second and puts out some specific form of heat, light, sound or mechanical motion as required.

Just what does this general view of the electrical system establish? Of specific processes—nothing; but at the transfer points we do begin to learn something. We know:

(a) that conventional fuels provide electrical power, so at least one of the conversions:

$$\left.\begin{array}{l}\text{thermal energy}\\ \text{mechanical energy}\end{array}\right\} \quad \text{electrical energy}$$

is possible;

(b) that the operation of domestic appliances demands that all of the conversions:

$$\text{electrical energy} \left\{\begin{array}{l}\text{mechanical energy}\\ \text{thermal energy}\\ \text{light}\\ \text{sound}\end{array}\right.$$

are possible;

(c) that different materials have varying ability to accept or to transmit electrical energy, the flow being able to travel long distances through metal wires yet unable to penetrate the plastic covering of domestic flex or to leak out into the air as it comes through the grid.

We could at this point return to our domestic appliances and look more deeply at them. We could open our switches and note that their action is just to introduce a small air gap in an otherwise complete loop of metal: we could note that a normal lamp consists of a fine wire which becomes extremely hot when connected to the supply system, while a fire has a slightly thicker wire and becomes less hot: we could open the vacuum cleaner and relate its wiring and construction to the forces which spin its fan. Among the complex of devices which fill the modern household we should certainly find enough material to define all the laws of electricity and magnetism.

But to do this would make our way too tortuous. We should be engulfed in detail as we looked for the underlying purpose of a design, because although the rules which describe electricity and magnetism are few in number, the variety of their applications in modern technology verges on the infinite. We should unnecessarily be putting ourselves in the position of the biologist, whose experimental system is defined and whose job is to explain it, when we are free to put ourselves in the role of the physicist who looks at simple things and builds complication only as far as his understanding will take him. Let us agree therefore to describe electrical phenomena in terms of a model whose sophistication will at all times be minimized, but which will have sufficient detail to account for the various effects which take our interest and the various devices we wish to understand.

How electricity moves about

Quite the best known feature of electricity must be its mobility. We have no hesitation in talking about electric currents, and we know electricity to be something which flows through the wiring of our houses, through the twistings of neon signs, through the storage cells and starter motors of our cars, even through the zig-zag of a lightning flash. Only when we begin to question what it is that flows, does general knowledge become less certain: is it electric charge, ions, electrons, atoms . . . ?

For our studies in electricity and magnetism, we shall make this the starting point, the need to identify a flow of electric charge. Our aim will be to specify the flow not just by a name, but as something which has recognizable properties through which it interacts with other things. From these properties we want to understand how a current may be initiated and how it can pass through such diverse media as copper wire, the vacuum of a television tube, or the acid of a battery, while other media, not greatly different, like plastic, glass and pure water, will scarcely permit it to flow at all.

Our study will be made from the start at the atomic and sub-atomic level.

2.1 Recognizing electric charge: what properties does it have?

In Unit 0, Modelling I, you encountered the atomic model used as an aide-memoire to synthesize relationships on the atomic scale. Throughout this course you will meet this model over and over again, each unit picking up, emphasizing and expanding particular aspects of it as they become relevant to the different subjects you study, and neglecting others of no immediate relevance to the problem in hand.

In these units on Electricity and magnetism we shall begin this process of elaboration. Here we shall fix our attention on the electrical aspects of our model, looking in detail at those features which contribute to our understanding of how charge moves, or may be made to move, while drawing only lightly on the patterns of atomic and molecular structure with which you have already become familiar. Our overall requirement of the model in these units will be that it leads us naturally to an appreciation of the general electrical properties of a variety of materials and ultimately that it shall account for their magnetic properties as well.

Few versions of the atomic model are able to ignore electrical charge, and that required for molecular structure was no exception. There, you encountered the idea that electrical charge is seated in the sub-atomic particles called protons and electrons. From the viewpoint of that model the properties required of these particles were simply that protons should repel protons, electrons should repel electrons, and that each should attract the other. An additional requirement was that proton and electron, when paired, should experience no net force in the vicinity of other unpaired protons or electrons (Figure 2).

It is through this need to annul forces that we come to build our conceptual models in terms of the single quantity, electrical charge, rather than by using the above force laws between protons and electrons. By supposing that protons and electrons have this one property, possessing it in equal amounts but in opposite senses, we come to a very succinct statement of our laws of force: like charges repel, unlike charges attract. Forces of this type, we refer to as *electrostatic forces*.

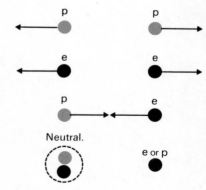

Figure 2

electrostatic forces

We need some further detail if we are to use this model of sub-atomic force. In an arbitrary choice we name the proton as having the positive charge, which leaves only the question of how the forces vary with distance to be specified. All intuition naturally suggests that the forces must diminish with separation and we, in our model, shall certainly suppose this to be so. The precise law which we shall assume, and one well supported by experiment, is that the forces vary inversely with the square of the separation (Figure 3).

Since electric charge in our model is just a convenient description of the forces between sub-atomic particles, an electric current must be a flow of particles having the property of electric charge. Because of our choice in assigning positive charge to protons what we choose to call a conventional large-scale electric current could correspond to a flow of protons in the direction of the current. Equally, because electrons are the charge opposites of the protons, it could also be a flow of electrons in a direction opposite to the current, or even some combination of electron and proton flow (Figure 4).

As we shall see later, electrons are the more mobile of the two electric species, and most currents are actually flows of electrons. This makes our choice of direction for the conventional current, which is no more than an historical accident dating from before the days of atomic science, a little inconvenient. It need have no effect on our reasoning, but it must be remembered. We can describe the force between two protons by the equation:

$$F = 2 \cdot 3 \times 10^{-28} \times \frac{1}{r^2},$$

where $2 \cdot 3 \times 10^{-28}$ is an experimental constant and r is the charge separation measured in metres. This is a very small force indeed. Even at a spacing of 3×10^{-10} m, roughly the atomic spacing for most solids and liquids, the force only rises to:

$$F = 2 \cdot 3 \times 10^{-28} \times \frac{1}{(3 \times 10^{-10})^2} \sim 2 \cdot 6 \ 10^{-9} \text{ newton.}$$

Between groups of protons or electrons the forces may be bigger. To describe this we use an equation which makes the total force proportional to the number of charges, n_1 and n_2, in each group:

$$F = 2 \cdot 3 \times 10^{-28} \times \frac{n_1 \times n_2}{r^2} \text{ newton.}$$

We elaborate this equation to show direction by taking the numbers as positive for protons and negative for electrons so that a positive force represents repulsion and a negative one attraction.

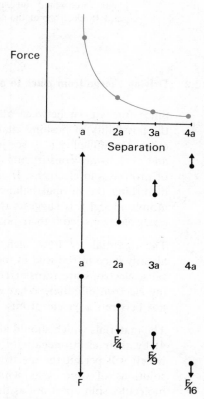

Figure 3

Conventional Current.

Electron Current.

Proton Current.

Mixed Current.
Electron. + Proton.

Figure 4

Exercises

What force will be exerted between a group of two electrons, and a group of three protons at a spacing of 6×10^{-10} m?

$$F = 2 \cdot 3 \times 10^{-28} \times \frac{(-2) \times (+3)}{(6 \times 10^{-10})^2}$$
$$\sim -3 \cdot 8 \times 10^{-9} \text{ newton.*}$$

The minus sign shows that the force is attractive.

* $\sim$ Means approximately equal to.

What force would you expect between two groups each containing 17 protons and 18 electrons at the approximate atomic spacing of 3×10^{-10} m?

We know that charges annul forces in pairs

$$F = 2{\cdot}3 \times 10^{-28} \times \frac{(17-18) \times (17-18)}{(3 \times 10^{-10})^2}$$
$$\sim 2{\cdot}6 \times 10^{-9} \times [(-1) \times (-1)]$$
$$= 2{\cdot}6 \times 10^{-9} \text{ newtons.}$$

The force is of repulsion and is exactly what would be found between single electrons at the same spacing.

2.2 Driving charge from place to place

Since forces exist between all charged particles we can easily recognize the possibility of pushing charge about merely by bringing other charges close. The difficulty is to see just *how* this can be done, for both electrons and protons are constituents of atoms, and atoms contain equal numbers of protons and electrons. If charge is to be moved we must find ways of disturbing the normal balance of electron and proton. In terms of our atomic model this suggests that means must be found to separate some outer electrons from their positive ion cores.

The essentials of how such a process may be driven chemically have already been introduced in the discussion of atomic bonding. In the ionic salts, electrons are transferred between atoms because of strongly differing electron affinities. What we need to do is contrive a similar transfer, not between adjacent atoms, but between pieces of bulk material.

The materials which should allow this transfer are already tabulated (Unit 4) by their electronegativity; the problem is to find a transfer medium which will permit the electron balance to be changed. We find it in the solutions of ionic salts which we call electrolytes. In such solutions molecules split apart not as their normal constituent atoms but as charged ions. A solution of common salt, for example, contains positively charged sodium ions Na^+ and negatively charged chlorine ions Cl^-, not just neutral sodium and chlorine atoms.

Through movement of ions like these, a transfer of charge may take place between materials of differing electronegativities suspended in the electrolyte. Devices which allow this kind of charge movement to take place on a continuous basis we call batteries.

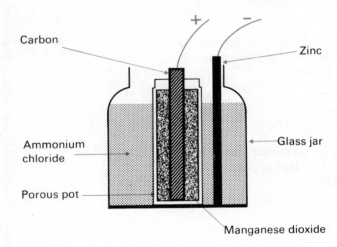

Figure 5(a).

Batteries

The dry battery in your torch or transitor radio is a commercial variant of what is called a Leclanché cell, Figure 5a. In such cells two different pieces of conducting material, called the electrodes, are set in contact with a chemical solution—the electrolyte. The electrodes of a Leclanché cell are of zinc and of carbon, and the electrolyte is a solution of ammonium chloride, $NH_4 Cl$, which yields ammonium NH_4^+ and chlorine Cl^- ions.

For its action, the cell depends on the differing electronegativities of carbon and zinc. Carbon is strongly electronegative, clinging to its electrons, while zinc, which is rather like calcium in Figure 53 of Unit 0, holds relatively weakly to its electrons. Now the tendency of zinc when placed in ammonium chloride is to dissolve. When zinc atoms do dissolve in this electrolyte they enter as doubly charged ions, Zn^{2+}, leaving two electrons behind in the body of the zinc. This process, of course, makes electrolyte positively charged and, in the Leclanché cell, leads to the release of ammonium ions, NH_4^+, at the carbon electrode where they capture electrons and leave the carbon positively charged. The overall result of zinc dissolution is thus to build up negative charge on the zinc and positive charge on the carbon.

A freshly dissolved and positively charged zinc ion lying close to the increasing negative charge of the zinc is of course attracted to it, and after a quite small amount of charge has built up these zinc ions are unable to leave the metal and the whole process stops. A similar effect occurs at the carbon. When a connection is made externally between the zinc and the carbon—perhaps through the bulb of a torch or the transistors of a radio receiver—electrons flow to the carbon from the zinc. This neutralizes the charge which has built up on both electrodes and therefore releases more zinc into solution. If the external connection is maintained, a continuous current will flow as the zinc steadily dissolves.

At this point difficulties arise. The principle of the reaction is fine: by going into solution a weakly electronegative material (zinc) gives up electrons though an interaction with a strongly electronegative material (carbon), but the overall process becomes inhibited because of detailed interactions at the surface of the carbon. When an ammonium ion NH_4^+ is neutralized by collecting an electron, it turns into ammonia, NH_3, which goes back into solution, and hydrogen, H_2, which does not. The hydrogen impedes the reaction of the cell and has to be removed. In finding ways to do this we go from simple scientific principles to the complications of technological adaptation.

In the old-fashioned Leclanché cell, which you may have seen in very old office telephone systems, manganese dioxide was packed in a porous pot around the central carbon rod. The effect of this was to absorb the hydrogen as it was released, in a reaction which turned it into water. The cell could thus provide a continuous current until its zinc electrode was completely eroded. The action of the manganese dioxide in capturing the hydrogen is called *depolarization*, and any material which behaves in this way is called a depolarizer.

depolarization

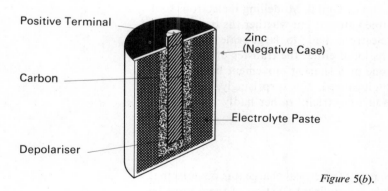

Positive Terminal

Zinc
(Negative Case)

Carbon

Electrolyte Paste

Depolariser

Figure 5(b).

In the modern adaptation of the Leclanché cell (the dry battery) the same ideas are used, but the components look different; it is not even obvious (Figure 5b), that an electrolyte is used. The carbon rod, as always, is central, but the zinc electrode becomes the case. Between these electrodes the electrolyte, of mixed zinc chloride and ammonium chloride, is a paste

11

stiffened with flour and starch and packed around a powder of manganese dioxide which is in contact with the carbon.

In these cells the principle of action is unchanged, but the passage of current now gradually thins the outer case. This is why dry batteries eventually become covered with an unpleasant white powder and why modern batteries often have an external steel jacket.

It is important to realize that battery action is not an exclusive thing confined to a few metals and electrolytes. Almost anything will do provided the output is not supposed to be constant and provided you do not expect it to light a torch.

Experimental interlude

Try making a battery. Damp paper between dissimilar coins, such as 2p and 10p, should give an output which can be read on your home kit meter.

If your cell does not work at first, a dash of salt sprinkled on the paper will probably help, or, alternatively, you can try a quite different electrolyte such as a slice of orange or apple.

See if you can find ways to combine two or more cells to obtain doubled or tripled outputs and look also for a form of connection which will give a near zero output.

Chemical means are not the only ones we have available for moving charge about. In your home kit you will find a solar cell. If you connect it to your meter and allow light to fall on it you can see the effect of a charge separation driven by the process called photoelectricity, in which electrons are ejected from material by energy absorbed in the form of light. Heat and mechanical force also can be used to separate or drive charge, through what are called the thermoelectric effect and the piezoelectric effect. Of all these processes the chemically motivated separation of charge gives by far the largest effects, but even this is small compared with the magnetic effect which we shall meet in Section 5.2.

2.3 Charge movement and its effects

We now have some idea of what is meant by electric charge and of how forces may be generated to drive it about. What we still lack is an understanding of how charge is able to move through materials in the concerted manner that we call a current.

In the materials described at the end of Unit 0, Modelling I, electrons took a major part in the bonding between atoms; but whether the bonding was ionic, covalent or dative the electrons had no large-scale freedom of movement. The various bonds required either the transfer of an electron from one atom to an adjacent one or a sharing movement between two, but they always held the electron localized. Not surprisingly, most of the compounds you considered conduct electricity rather badly.

2.3.1 Charge flow in metals

If a material is to permit the free flow of electrical charge it is obvious that all its electrons cannot be bound. A significant fraction of them must be able to take part in a more general movement not tied to specific locations within the material.

The atomic model which gives the greatest freedom to electrons is that appropriate to metals. Materials such as potassium, lithium, copper, cadmium and zinc enter into chemical bonds by giving up electrons—

either completely, in ionic bonds, or partly, in covalent bonds. When the atoms of such materials are assembled among themselves to form either pure metals or alloys they have no convenient and adjacent dump (such as an electron-hungry chlorine or sulphur atom) into which a valence electron can be put and with which a localized bond can be formed. Instead, the atoms form an overall alliance, pooling their valence electrons throughout the whole material. The valence electrons are thus unlocalized: they spread about through the metal, moving around between the positive ion cores, from which they have separated, like the atoms of a gas.

Charge movement in a material like this is easy to understand. The positive ion cores separate as best they can, forming a regular array from which they cannot escape. The light and mobile valence electrons also spread out into a uniform distribution, but this distribution is a dynamic one with individual electrons moving first in one direction and then another as they bounce off each other in mutual repulsion (Figure 6).

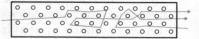

Figure 6

With such a material, the addition of electrons at any point produces an overall repulsion on all other electrons so that their random bouncing motion becomes a drift which rapidly smoothes away the excess of concentration. If electrons are added at one point and removed at another there is a drifting motion from input to output point and we say that a current flows.

Examples

We can get some idea of the power of these levelling forces if we use our example of the forces between electron and proton. A rod of copper 1 cm² in cross-section and 10 cm long weighs about 90 g. It contains about 8.5×10^{23} atoms and about 2.5×10^{25} electrons and protons. Suppose that somehow an electron excess of just 1 part per million was caused in one half of the rod, with a corresponding deficit in the other. Let us calculate the forces available to bring back the normal uniform electron distribution.

Our interest is not in exact answers—we just want orders of magnitude so we shall suppose that the electrons can be treated as if they are concentrated at the mid-points of each half of the bar. The problem is therefore equivalent to finding the force between a group of $10^{-6} \times 2.5 \ 10^{25}$ electrons 5 cm distant from a similar group of protons. We know this force to be:

$$F = 2.3 \ 10^{-28} \times \frac{(-2.5 \ 10^{19}) \times (+2.5 \ 10^{19})}{(5 \ 10^{-2})^2} = -6 \ 10^{13} \text{ newton.}$$

This time we have a force of attraction which is very large. As it requires only about 10^7 newton to produce severe deformation in a copper rod of this section we can be quite sure that the electrons will move so as to establish a more uniform distribution. If the electrons were not free to move, they would simply collapse the rod.

This example shows that when we discuss changes in electron concentration we mean changes on a scale beside which one part per million is huge.

Exercise

Do you think the force on just one electron in the above example is large enough to be a measurable quantity?

The force on the group of electrons is the total of the forces acting on each of its members. The force per electron is therefore $6 \times 10^{13} \times (2.5 \ 10^{19})^{-1} = 2.3 \ 10^{-6}$ newton. This is about the weight of a mass of 1/4000 g; this is just below what you might feel with your finger and a force of ridiculous dimensions for the atomic scale.

2.3.2 Description on a larger scale

The picture we have used of unconstrained electrons bumping their way through a metal invites a quantitative description in terms of electron speeds, the distance between collisions and the number of electrons free to move. Such a description can indeed be assembled, but for most purposes it is too detailed. Often we want just to be able to say that under some given conditions the rate at which charge flows through a material will have a particular value. Our next task is to find how to state these conditions and to describe the rate of charge flow.

To make this clear, think of another case of flow that we know well enough to treat as intuitively obvious; think of a pipe with water flowing through it from a small reservoir as in Figure 7. With water, we have no difficulty in forming an adequate description of the flow. Water flows through the pipe when a pressure is available to drive it, and it flows faster the greater the available pressure. An adequate description of the flow therefore consists in being able to make a set of statements like: when the pressure difference between the ends of the pipe is 100 N/m² the water will flow at a rate of 50 1/s.

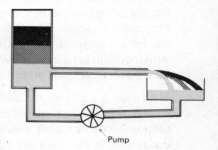

Figure 7

For our electrical description the pipe is the mesh of positive ions which provide the structure of the metal, the water represents the electrons and the pump is the battery which feeds electrons to the metal at one point and collects an equal number at another. Since the analogue of the rate of water flow is clearly the number of electrons entering or leaving the metal in one second, only one concept is needed to make our analogy complete; we must find the electrical equivalent of pressure.

Now we know that when electrons flow through a conductor they do so in response to forces caused by minute variations of electron concentration. If an opposite, and slightly larger force could be applied to an individual electron, it would drift its way backwards against the normal flow, making its irregular way from the normal output point to the normal input point. In making an electron follow such a reverse path, work (see the unit on Mechanics) would need to be done. But this work is effectively stored by the electron as potential energy, for if the electron is released it will travel back through the material with the electrostatic forces pushing it all the way. It is as if an electron at the input point is at the top of a hill ready to slide down through the material. The potential energy which a charge loses as it moves in this way is what we use as our measure of electrical pressure.

At this point history lets us down. It would be natural to use the potential energy held by just one electron as our measure of electrical pressure. We should expect therefore to express our unit of potential as an energy store per unit of charge and therefore as so many joules per electron. But the subject did not develop this way. The big scale experiments were done first and the atomic model we have developed came later. The result is that we use a larger measure of charge and consequently a different unit of potential.

The charge unit we use is a huge one called the *coulomb*. This is about $6 \cdot 3 \times 10^{18}$ times the charge carried by one proton, so that the proton charge expressed in terms of this unit is $1 \cdot 6 \ 10^{-19}$ coulombs and the electron charge is $-1 \cdot 6 \ 10^{-19}$ coulombs.

coulomb

We measure potential as so many joules of energy per coulomb of charge. We commonly call this unit of potential—the joule per coulomb—the *volt*.

volt

The way we describe potential is to say that there is a potential difference of 1 volt between two points if an energy of 1 joule is expended in shifting

14

1 coulomb of charge from the lower potential point to the higher. The way we use this potential is to say that a voltage, or potential difference, drives charge through a conductor.

To exploit the idea of potential we need yet another unit of electricity. Since the effect of a voltage is to drive charge through a conducting material, we can only express the result of applying a voltage if we have a unit describing the rate at which charge flows. On the atomic scale we would of course work in terms of the number of electrons passing some point per second. On the larger scale we must use the coulomb per second to describe flow rates. This unit is better known as the *ampere* or amp.

ampere

With this final unit, we complete our analogy with water flow. We can now express the parallel to our statements about pressure and fluid flow by statements of the form: When the potential difference between the ends of this conductor is 5 volts, electrical charge will flow through it at a rate of 3 amperes.

2.3.3 Some effects of charge movement on different materials

In effect we have now studied the charge-moving properties of three kinds of material. We have considered metals in some detail, have rather dismissed the bound electron materials (the insulators) as of no great interest, and have made fleeting mention of charge motion through the electrolytes of batteries. The metal and the insulator represent extremes. Between them we have a whole range of materials of varying conductivity whose individual electrical properties have far more interest than just the ability to pass charge.

In this section we shall return to our atomic model and, holding to the problem of charge flow as our theme, we shall explore some of these other properties and some of the applications which give them industrial importance. With our model at this stage we shall begin to recognize in it many of the electrical phenomena lying within our common experience.

We shall begin by returning to metals. The most obvious of all electrical effects is that of heating. Let us see just how our atomic model can be made to account for this.

In a metal carrying a current, valence electrons will make their way from positions of high potential energy—remember that this means low voltage since the electron charge is negative—to places where their potential energy is low. Between collisions an electron will be accelerated by electrostatic forces and drift with them, but after each collision its velocity will be essentially random. The forces thus do not give the electron an overall increase in kinetic energy. Instead they serve to increase the overall random kinetic energy of everything with which the electron collides. As you may know, and as you will find later in the unit on Energy conversion, randomly directed kinetic energy is just heat. The electron as it moves thus releases its potential energy by heating the metal as a whole.

Not all media present as many obstacles to electron motion as does a metal. A vacuum is a fine example of a medium providing virtually no obstacle to electron motion; in it, electrons may accelerate continuously between electrodes at different potentials, gaining speed and hence kinetic energy as they lose potential energy. We will begin our survey of other materials by considering the conduction of charge through gases, taking as our first consideration the extreme case of a vacuum.

2.3.4 Charge movement in gases

A gas is a mixture in which molecules move about at high speed interacting with each other only through collisions. At normal temperatures and pressures, the collisions are very frequent and the spacing between them is small—10^{-6}m say—but if the pressure is reduced, the mean distance between collisions becomes larger. When the spacing between collisions approaches the distance across the enclosing low-pressure chamber, we begin to describe the cleared space as a vacuum.

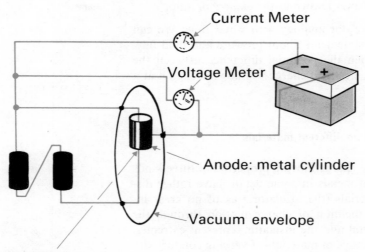

Figure 8

It may seem that a vacuum is a hopeless medium in which to describe the movement of charge because it has no substance to support the charge, but in fact charged particles can be put into a vacuum and their movement within it may be directed with some precision.

In the old-fashioned vacuum tube diode (Figure 8), electrons are emitted by a hot tungsten wire, called the *cathode* (at high temperatures in this metal, a few valence electrons near the surface may gain enough kinetic energy in collisions to escape the binding forces which hold them to the metal) and be accelerated through the vacuum to a plate, the *anode*, held at a high voltage relative to the wire. As in a metal, the electrons gain energy by moving to a position where their potential is lower, but in the absence of collisions this is done by acceleration and the energy does not appear as heat in the vacuum: heat does appear however in the anode which the electrons strike at high velocity, turning their considerable kinetic energy into random motion of the atoms in the anode.

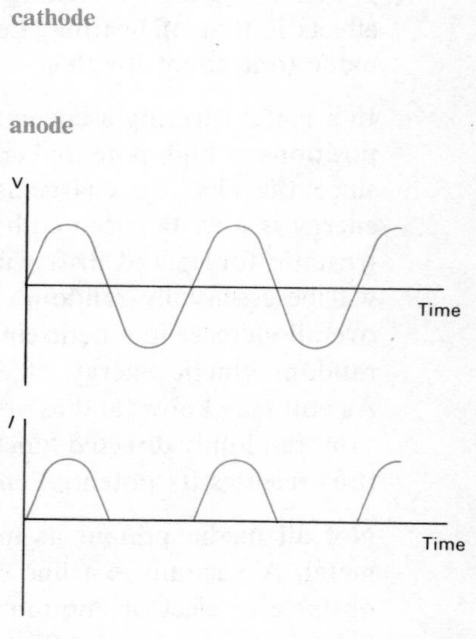

Diodes of this type, once very common, are used to rectify currents. By their nature, they can pass currents in just one direction because only the filament can emit electrons. Therefore if a varying voltage is applied between cathode and anode, current will flow only when the anode is at a positive voltage relative to the cathode (Figure 9).

The television tube

A more sophisticated use of the vacuum to carry an electric current is the oscilloscope or the television tube. This uses an electron gun, which is effectively a diode with a hole in the anode, to form a narrow beam of electrons along the axis of the tube (Figure 10). Fluorescent material on the end of the tube then emits light when struck by the electron beam. The precise design of a gun to provide a good fine beam demands rather more than just a diode, but it is on the behaviour of the beam after leaving the gun that we shall concentrate now.

cathode

anode

Figure 9

After leaving the gun the electrons are in a region of unchanging potential so they do not accelerate but move at constant velocity towards the screen. If a pair of plates is placed to either side of the beam and a potential difference is maintained across them, the electrons will be able to reduce potential energy by moving towards the plate at the higher voltage. They will accelerate towards this plate, but will not normally reach it because their high velocity towards the screen takes them past it before they have been deflected sufficiently to reach it. What is observed instead, is that the beam is deflected across the fluorescent screen. By using two pairs of plates arranged perpendicularly the beam may be deflected to any part of the screen.

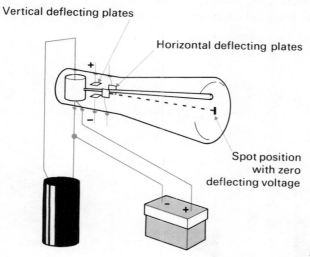

Vertical deflecting plates

Horizontal deflecting plates

Spot position with zero deflecting voltage

Figure 10

Elaborations of this type of device and other similar ones, give us the domestic television tube and the laboratory oscilloscope. In the television tube a pattern of voltages is fed to the plates to make the beam scan the screen in a progressive manner, the picture being formed by changing the intensity of the beam to give light and dark as required. In the oscilloscope the emphasis is on the measurement of varying voltages. A ramp voltage, which rises at a steady rate then drops quickly to its starting value, is used to sweep the beam horizontally across the screen and bring it back again. An unknown varying voltage is then used to drive the vertical deflection plates so that a trace, showing how the voltage varies with time, appears on the screen.

The oscilloscope is nowadays the workhorse of technological measurement. Its commercial form is vastly more sophisticated than that described above: a typical specification would require the beam to traverse a 10 cm screen in times varying to well below 10^{-6} second (1 microsecond) and to provide a deflection of 1 cm for a signal of 10^{-3} volts (1 millivolt). The versatility of the instrument is often enhanced by the provision of two beams, to display independent signals in synchronism; a trigger, to commence the sweep on receipt of a pre-selected signal; and a memory device, to hold the display of a signal long after it has been recorded.

In the diode and the cathode ray tube we have been concerned with how electrons move in the vacuum between electrodes held at different potentials. Similar considerations should apply when other kinds of charged particle move between electrodes, even when the surrounding medium is a gas at rather higher pressure. This is a matter of considerable technological importance so we shall look at it more closely.

Any charged particle in a region between two conductors at different potentials will be able to reduce its potential energy by moving towards one of them. Positively charged particles will therefore tend to move towards the conductor at the lower voltage while negatively charged particles will tend towards the conductor at the higher one. The shape of the conductors is of no particular importance; charged particles will tend to drift, even after collisions with atoms of a surrounding gas, towards the conductor at the appropriate potential.

Electrostatic painting

A number of industrial processes use this effect to direct the movement of small charged particles through a gas. One such process is electrostatic painting, in which a large potential difference is maintained between a spray gun and the work piece. The paint is charged as it leaves the gun and the paint globules are then driven towards the workpiece, and they are attracted towards every surface to give complete coverage of even the most elaborate shapes.

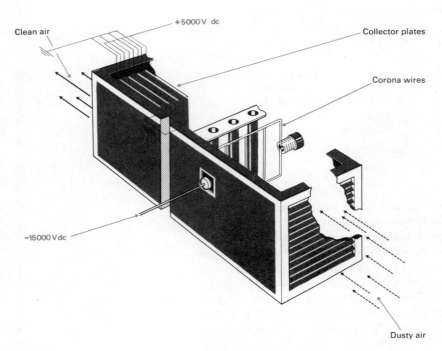

Figure 11

Dust precipitators

An application of electrostatic forces of major importance is the electrostatic precipitator, a device which is used to extract dust or small particles from air or furnace exhaust gases flowing through it. On entry to the precipitator (Figure 11) gas first passes a row of wires held at a negative potential of about 15 000 volts and then past a row, set behind and between the wires, and maintained at earth potential. The very large potential difference between rods and wires causes ionization around the wires in what is called a corona discharge and has the effect of giving negative charge to the dust particles in the gas. By passing the gas with its charged particles between plates alternately earthed and at a positive potential of 5 000 volts the dust particles are driven to the positive plates and adhere in clusters which are released by shaking when they are heavy enough to fall. This process of precipitating dust is not a small-scale one. In a large modern power station the precipitators may handle about 100 tons of dirt per hour while missing about one half of a ton, mainly in the form of very small particles which drift extremely slowly across the gas flow.

18

So far we have avoided what may seem the simplest case of charge flow through a gas, flow through a gas at normal temperature and pressure un-influenced by the injection of extraneous particles like electrons, paint globules or dust. The reason is that the mechanism of charge flow in a gas is more complicated than might at first be expected. All the materials and processes with which we have dealt so far have had one common and sim-plifying characteristic: they have relied for the movement of charge on the mobility of a single type of particle. In the metal, we considered the motion of electrons, and in a vacuum, it was again the motion of electrons or the motion of heavier particles with a single type of charge. Now that we come to consider the passage of charge through pure gases, ionic solids and elec-trolytes, the process of conduction is complicated by the simultaneous motion of two species of charged particle, for in these materials there is no fixed reference grid like the positive ion mesh of the solid, and positive ions are just as capable of movement as negative ones. Gases in their nor-mal state scarcely conduct electricity at all. When a modest potential diff-erence is maintained across a pair of plates separated by some millimetres of gas, such charge as passes between them is entirely due to chance ioniza-tion. This can be caused by cosmic ray showers, adjacent X-rays or radio-active material, all of which are able to eject electrons from normal un-charged gas atoms.

When large voltages are used, appreciable conduction can occur. It arises from breakdown of the normal atomic structure of the gas which is changed from a collection of electrically neutral molecules into a mixture of ionized atoms or molecules and free electrons, together with normal atoms and molecules. We call this reactive mixture a *plasma*.

plasma

Suppose that in a normal gas experiencing a large electric field a chance event causes the ionization of a single atom. The ion, which as an atom experienced no electrical force, will immediately be accelerated towards the cathode, while the freed electron accelerates towards the anode. As they move, both particles will collide with other gas atoms and exchange momentum and kinetic energy with them.

If the charged particles travel far enough between collisions they may gain enough energy to strike electrons from other atoms during collision. If this happens, there will be a rapid spread of ionization throughout the whole column of gas and a steady current will begin to flow. This current will be due to a mixture of ionic and electronic motion. The neon lamps used for illuminated signs pass currents of this sort, the light emission aris-ing because the buffeting which the atoms receive in collision gives them energy which they are able to release as light.

2.3.5 Charge movement in ionic materials

In ionic solids there is relatively little conductivity at normal temperatures. Each ion has a specific set of electrons associated with it and charge motion is scarcely possible. At high temperatures the passage of an electric current becomes relatively easy. What happens is not that electrons are freed from the ions, but that the ions themselves begin to jump about in the crystal lattice of the solid with a random diffusive action which an electric field will turn into a drifting movement. Depending on the solid, either ion, or both, may move. At temperatures such that the salts are liquid the move-ment naturally becomes easier, and electric charge may flow more freely.

The characteristic of these salts is that any movement of charge is asso-ciated with a movement of actual material. As a consequence some very reactive metals, which defy separation from their salts by normal chemical means, are commonly extracted in this way. We may take aluminium as and example (Figure 12). First the ore, bauxite, is converted to pure aluminium hydroxide. This is then roasted (calcined is the technical term) in a rotary kiln at a temperature of 1 200°C, a process which removes all water and converts the hydroxide into the oxide of aluminium (Al_2O_3). The oxide, alumina, has a very high melting point and so is itself unsuitable for use in a process involving atomic movement. It will however,

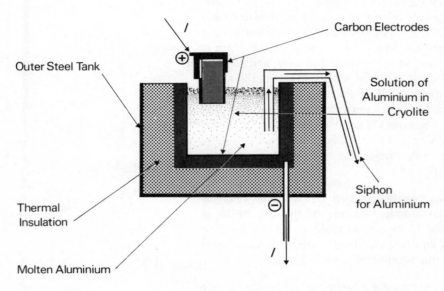

Figure 12

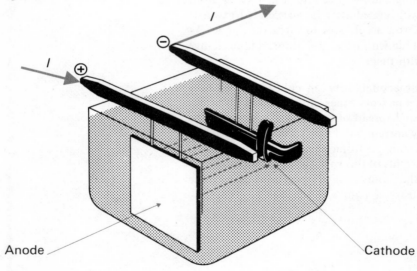

Figure 13

dissolve at more modest temperatures, 1 000°C, in a material called cryolite (AlF_3: $3NaF$) and in the commercial process it is from this hot solution that aluminium is separated. Treatment takes place in large cells which pass about 100 000 amps between carbon electrodes. Although aluminium is the metal finally liberated, it is a movement of positively charged sodium ions, Na^+, from the cryolite which supports the electric current. These ions drift towards the negative electrode, or *cathode*, where they gain an electron and then displace atoms of aluminium from the cryolite. At the temperatures involved, this aluminium remains molten and it accumulates in the lower part of the cell, being removed periodically as required. The process therefore is one in which

cathode

a movement of electrical charge is accompanied by a movement of material.

You have already met motion of this sort when we discussed the battery, where the active material was a solution of an ionic salt, or electrolyte. In that case, an inherent tendency of certain chemicals to redistribute themselves was used to drive an electrical current in an external circuit. By contrast in the present case, a current driven through the material is used to produce a chemical separation. Electrolysis is the name of this process. In it, ions liberated at the electrodes are deposited as solid material, or released as gas, or take part in chemical interactions with electrode or electrolyte. The simplest case occurs when the metallic ion of the electrolyte is of the same material as the anode or positive terminal. Passage of a current then transfers anode material and deposits it on the cathode. A wide variety of metals may be deposited in this way (Figure 13).

Electroplating

This process is of considerable importance industrially. In galvanizing it is used to deposit a rust-preventive zinc coating on iron components, a zinc sheet being used as the anode. Another application is in the purification of copper; in this, copper from a block of impure copper used as the anode is deposited on a rod of pure copper used as cathode. Metallic impurities may be passed by this process, but non-metallic materials cannot transfer into the solution and are effectively excluded from the deposit.

Electrolysis of water

When inert electrodes are used, perhaps platinum or carbon, and an acid, hydrochloric say, replaces the salt solution, a current will still pass. In this case the ions are of hydrogen H^+ and chlorine Cl^-, and their movement causes a release of hydrogen at the cathode (Figure 14).

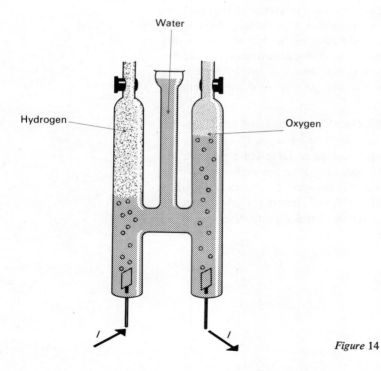

Water

Hydrogen

Oxygen

I

I

Figure 14

At the anode, however, chlorine is not released; instead, the chlorine ions react with the water taking its hydrogen and releasing oxygen. The net effect is that the water is separated into its constituent parts, the acid being left in solution.

21

Fuel cells

A particularly interesting feature of this reaction is that the hydrogen and oxygen can recombine with a release of energy: together they constitute a fuel. In suitable situations this energy may be released directly in electrical form, effectively by reversing the electrolysis process. We call such an arrangement a fuel cell.

One arrangement for such a cell uses electrodes of compressed nickel powder through which the gases pass into a cooled electrolyte. The hydrogen atoms each give up an electron to their electrode, which thus becomes the cathode, and then pass as ions to the anode where they collect an electron and combine with oxygen. To be efficient these cells must work at an elevated temperature, 250°C, and none are yet in widespread use. They are, at the moment, among the technological exotica of spacecraft but could at some time drive humbler vehicles.

2.4   Résumé

This has been a descriptive section and it has covered a lot of ground. In it, our central aim has been to establish a relationship between the small and the big, to show that large-scale electrical phenomena may be described elegantly and succinctly in terms of a simple atomic model.

Our train of argument led us to identify electric charge in terms of forces exerted between sub-atomic particles called protons and electrons, and suggested that charge separation and accumulation could provide driving forces of the type needed to account for the concerted charge movement we call current. Elaboration of the atomic model then allowed us to appreciate some of the differences that materials show in their ability to conduct charge and led us to recognize in the model some of the better known large-scale effects.

Keeping to the theme that we must account for the overall charge-carrying properties of a range of media, our investigation of metals, ionic solids, electrolytes and gases served to introduce some of the domestic and industrial devices whose operation we questioned in our introductory section. Our account of conductivity in these media thus contained references to batteries, TV tubes, precipitators and the processes of metal refining, electrolysis and electroplating.

Because of our need in this section to develop the links of our large- and small-scale relationship we have skimped—purposely—the harder, the more definite and numerical relationships of large-scale electrical descriptions. We have done no more than introduce the main units of description, the coulomb, the volt and the ampere. In the sections which follow we shall begin to redress this balance, turning gradually to this form of description as a quick, convenient and quantitative way of representing many electrical phenomena.

Section 3

A review of electrical and magnetic effects

While atomic events underlie all the effects we know as electrical, description in such terms is too cumbersome if we want just to refer to the performance of a lamp or a battery or an electric motor. For domestic purposes, and for the majority of electrical engineering purposes, the large-scale type of description which we introduced in our discussion of the coulomb, volt and ampere is all that is required.

Surprisingly perhaps, the number of rules we must learn for our description of electrical phenomena is small. The great variety, both in size and in purpose, that we meet in domestic and industrial devices arises less from a multiplicity of scientific laws than from the application of a small number of laws in a wide range of combinations. We shall therefore devote this section to a brief review, in schematic and pictorial form, of the laws of electricity and magnetism with which we shall mainly be concerned. Equipped with a knowledge of these laws, which represent the experimentally backed scientific description of what is meant by electricity and magnetism, we shall move to our main business which is to show how intelligent application of these rules provides the basis for modern electromagnetic technology.

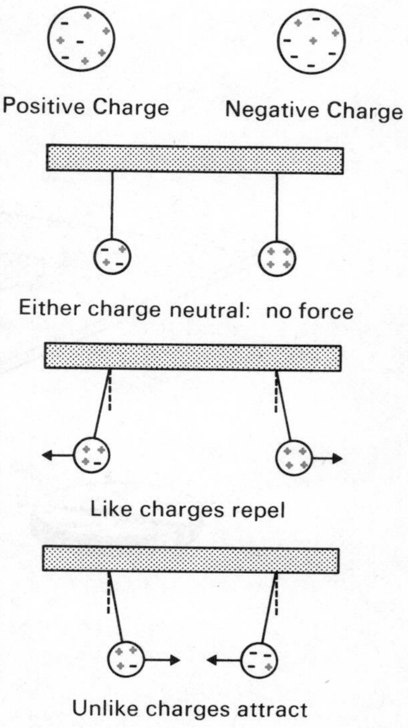

3.1 Six phenomena

3.1.1 Electrostatic forces

The first of our large-scale laws is already familiar. For large bodies as for small, electrostatic forces have effect and like charges repel while unlike attract (Figure 15). Elaborating this idea, we associate a surface of high potential with the tendency to attract negative charge and repel positive charge, while a surface of low potential attracts positive charge and repels negative charge (Figure 16).

3.1.2 Currents

Electric currents flow because small forces on electrons or ions cause them to drift through materials. The large-scale statement of this effect is that a potential difference must be maintained if a current is to flow continuously through a conductor. Conversely, to maintain a potential difference across a conductor a current must flow. Irrespective of the type of conductor, current always implies a difference of voltage, and voltage difference implies current. In terms of the fluid analogy, this rule requires that a driving pressure will always cause fluid flow through a pipe, and equally that flow through a pipe will take place only when there is a driving pressure.

3.1.3 Magnetic forces

A completely separate set of phenomena are involved in our third rule. Magnets have quite well-known properties: they attract iron, they attract or repel each other when brought close, and in certain shapes (notably long thin ones) they will swing, when freely suspended, to point roughly along a north-south line.

Positive Charge   Negative Charge

Either charge neutral: no force

Like charges repel

Unlike charges attract

Figure 15

Figure 16

When we refer to a magnet we commonly use this directional property to distinguish between its ends. We call the end which tends to point north the north pole or the north-seeking pole of the magnet, and the other end its south or south-seeking pole. The forces which act between magnets can very readily be described in terms of forces which act between the poles and if this is done the law of force becomes a very simple one for, just as with electric charge, like poles repel and unlike attract.

Because of the forces between magnets, a compass placed close to the south pole of a large magnet will rotate until its north pole points directly towards the magnet. Near the other pole, the south pole will point towards the magnet. At other positions the compass will take up intermediate directions, changing progressively as it is moved from point to point.

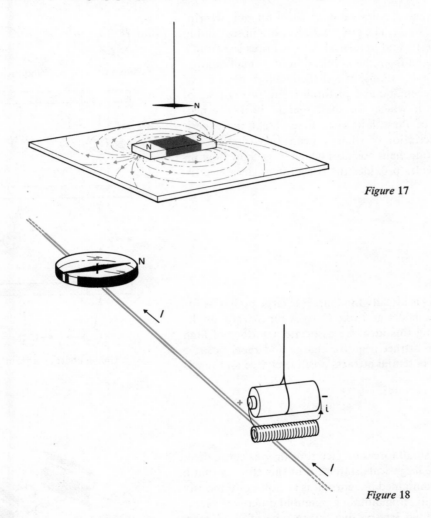

Figure 17

Figure 18

Maps of compass direction can be drawn around a magnet, as in Figure 17, with the lines at each point having their tangents in the orientation of the compass at that point. We describe this situation by saying that around a magnet there is a zone of influence which we call a *magnetic field*. At each point this field has a direction which is indicated by the north-seeking pole of a compass and a strength which, for the moment, we shall take as a measure of the tendency of the compass needle to return to its normal position if displaced.

magnetic field

3.1.4 Electromagnetism

When a compass needle is placed near a conductor, a wire say, through which a current is flowing, a remarkable unification of the effects of elec-

24

tricity and magnetism can be observed. The compass needle will be diverted from its normal north-south orientation and will tend to point in a direction perpendicular both to the wire and to a line joining the wire to the compass (Figure 18). A similar effect is observed using a small battery driven coil. In both cases reversal of the current direction will cause a reversal of orientation.

Exactly as with a magnet, a map of compass directions may be constructed around a conductor carrying a current. Depending on the shape of the conductor, the map may be complex or simple, but the region around the conductor will always show the characteristics of what we have already called a magnetic field. This, therefore, is our next rule: electrical currents produce magnetic fields in the region surrounding them.

3.1.5 Electromagnetic force

Since magnets and conductors carrying currents influence the space around them in a similar way, our next effect is hardly surprising. Just as magnets may attract or repel each other, so a wire carrying a current will experience a force if it passes close to a magnet or another wire carrying a current. In any region where there is magnetic field a current-carrying wire may experience a force. Figure 19 (a and b) shows two examples of this sort of force. We call forces of this type *electromagnetic forces*.

electromagnetic forces

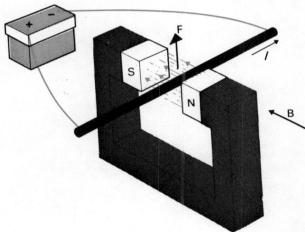

Figure 19(a)

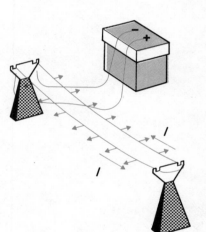

Figure 19(b)

3.1.6 Electromagnetic induction

The last effect we shall consider is rather like the inverse to the electromagnetic force. When a closed loop of a conducting material is placed in a magnetic field any change in the field will tend to produce a flow of current in the loop. The type of change which produces the greatest currents (Figure 20) is that in which the map of the magnetic field intersects the conducting loop and is then either removed or reversed. We refer to this effect, in which changing magnetic fields produce currents, as *electromagnetic induction*.

electromagnetic induction

3.2 The use of electromagnetic effects

Between them these six effects define all we need to know about large-scale electricity and magnetism. To be sure, we must define some of them much more carefully before they can be of use to us, and there are, of course, many other electrical phenomena we have not considered, but within the bounds of these effects we can pick out much of what was required in our original survey of electrical devices. Clearly the force acting on a wire carrying a current through a magnetic field can be the basis of a motor (the electrical to mechanical energy convertor of Section 1) if we have but the wit to manipulate field and force correctly. Equally, the reverse conversion should be possible by using mechanical forces to change a magnetic field (say, by spinning a magnet) and picking up the electric energy by holding a conducting loop in a region of continuously changing field.

In the sections which follow we shall take up the different phenomena reviewed here, showing how each of them is of consequence in our common technology and is not just a curiosity of the experimental laboratory.

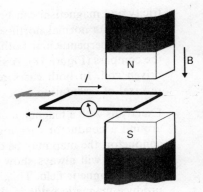

Figure 20

Section 4

Electric currents

When we considered the flow of charge through conductors in Section 2 our concern was mainly with the details of the conduction process, with the number and the mobility of charged particles. In this section the emphasis will be quite different. We shall again be concerned with charge flow, but this time we shall concentrate on formalizing the relationship between the available driving force—the potential difference—and the current which flows, largely ignoring the detailed mechanism of charge movement.

Our primary objective will be to manipulate statements of the type: when a potential difference of so many volts is applied to a given conductor, charge will flow through it at a rate of so many amperes. We shall develop a form of description which allows this sort of statement to be made more compactly and invites extension, allowing us to deal with the relationship of current with voltage—the I, V relationship—for groups of conductors in terms of known relationships for individuals.

4.1 Circuits

To come to terms with the idea of a current flowing through a conductor, consider first the arrangement of Figure 21. A number of conducting articles have been connected to each terminal of a battery, the connections being made either with metal wires or simply by touching the conductors to each other. Now this arrangement provides no continuous path by which current may flow from one terminal to the other, and in the absence of a current we know that no potential difference can exist in a conductor. The group of conductors attached to each terminal must therefore all be at the same potential.

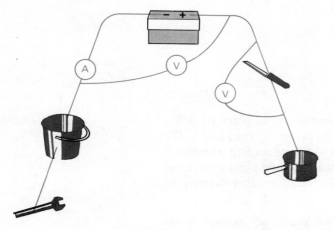

Figure 21

We can check this experimentally. If we connect a voltmeter (which we can regard as indicating potential difference by the deflection of a spot on an oscilloscope) across the knife or the pail it must show a zero reading. A connection made from one side of the battery to the other will show a voltage, however, for the function of a battery is, by chemical action, to generate just this difference of potential. Clearly we must expect to read the same voltage for all possible connections between points on one side of the battery to points on the other.

We have not yet considered the question of measuring current, but conceptually at least, one might think of an ammeter as a device to display the force produced by the current and magnet arrangement of Figure 19. If such a meter is included in our circuit along the line of conductors as shown, it can, for this arrangement, give only one reading—zero.

With an additional conductor added to complete the loop of conducting material between the battery terminals, as in Figure 22, the situation is completely changed. A potential difference now exists along a continuous conducting path; a current *must* flow. We must expect the ammeter to show a reading, and the voltmeter to give a value when connected across any conductor in the loop. With the voltmeter connected from one battery terminal to a series of points progressively further away around the conductor loop, an increasing set of voltages will be found, reaching their highest level at the other battery terminal. Thus the sum of the voltages across the conductors equals the voltage across the battery terminals.

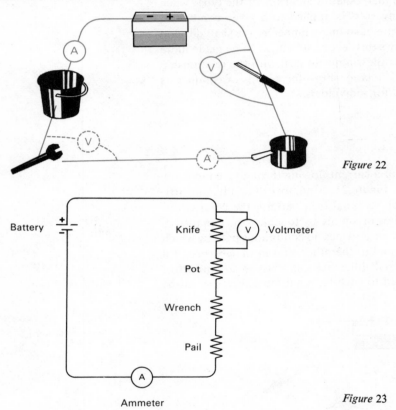

Figure 22

Figure 23

This might have been expected, since only one source of potential difference drives the flow, and it must be distributed across the conductors. The ammeter, in contrast to the voltmeter, gives the same reading wherever it is connected in the loop. Were this not so, charge would build up at some point in the circuit, and we know this to be forbidden because electrostatic forces always act to disperse charge.

Our simple circuit exemplifies a rule which was previously implied by the water flow analogy we used to illustrate flow on the atomic scale. A potential difference must be maintained if a current is to flow continuously through a conductor: to maintain a potential difference across a conductor a continuous current must be supplied.

A schematic representation of the electrical circuit we have been considering is shown in Figure 23. This is yet another example of a model; in this case the various household items have all been represented formally by zig-zag lines to emphasize that, for the purposes of our circuit at least, they all have a common property, the ability to allow a flow of charge when subjected to a potential difference. The difference in position of voltmeter and ammeter in the circuit is worth noting. The voltmeter always has the position of a bystander: it lies outside the main loop of a circuit and samples it by being connected across its different elements. The ammeter lies within the main path of the circuit and transmits the full current being measured. We shall see later that the difference between these instruments lies more in their use than in their design: the meter of your home kit is able to perform either measurement just by operation of a switch.

4.2 Resistance

This brings us to the central problem of this section, the development of a succinct form of description which will allow us to specify the current–voltage, or I, V, relationship not just for single conductors, but for groups of conductors connected in arbitrary fashion. We want to go rather further than this, to find ways of specifying current and voltage for each element of an array of conductors when some arbitrary current or voltage is supplied at any two of its junction points.

4.2.1 Electrical resistance of a single element

We begin with a single conductor and look for a short way of describing its current–voltage relationship. We do this by introducing the idea of electrical resistance.

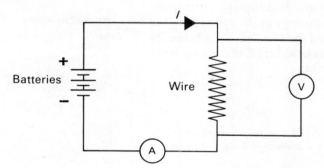

Figure 24

If a wire is connected between the terminals of a battery a voltage will be maintained across it and a current will flow through it. By using two, three or more batteries connected as in Figure 24 with positive terminal to negative terminal, larger driving voltages will be produced and larger currents will flow. A graph of the measured current I against the applied voltage V is likely to take the form of the full line in Figure 25.

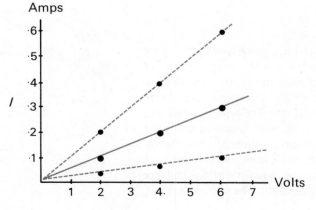

Figure 25

The use of a similar but thicker wire will give a generally increased current as in the upper line, while a thinner wire will give lower currents as in the lower one. In the same way, experiments with long wires will give smaller currents while shorter wires will give increased ones. The overall pattern in fact is identical with what you would expect from a series of narrow and wide, long and short pipes with water flowing through them.

For all cases it will be found that the relationship between current and voltage takes the form of a straight line over a reasonably wide range. The line must always pass back through the one experimental point we can be certain of, namely the origin, because no current can flow if there is no voltage. Within the region of this linear relationship, a set of measured current and voltage values (I_1, V_1), (I_2, V_2), (I_3, V_3), etc., taken from a given conductor all fit the equation:

$$\frac{V_1}{I_1} = \frac{V_2}{I_2} = \frac{V_3}{I_3} = \text{constant} = R$$

We call circuit elements which satisfy this equation, *linear elements*. As we shall see, such elements are very convenient for circuit calculation. A knowledge of any one pair of values for current and voltage allows the constant R appropriate to the conductor to be calculated. All other possible relationships can then be calculated using: $V = IR$.

This law, that current is proportional to the applied voltage, we call *Ohm's law* and it applies to a wide range of conductors held at constant temperature. It must be emphasized that the value of the constant R describes a property of the conductor. It has the significance that when R is large a given potential difference produces only a small current, and when it is small a relatively large current is produced. The constant can be regarded as describing a tendency of the conductor to inhibit the flow of current, and this is why we refer to it as the *electrical resistance* of the conductor. The unit in which resistance is measured is called the *ohm*, one ohm being the electrical resistance of a conductor which passes a current of one amp when subjected to a potential difference of one volt.

Exercises

The electrical resistance of a piece of wire is 4 ohms per metre of length. What voltage must be supplied to drive a current of 2 amps through a 3 metre length?

A torch bulb for use with a 6 volt battery is marked as passing a current of 0·5 amps. What is its electrical resistance?

Most metal objects, and many others, show an impressive tendency to obey Ohm's law. It is tempting to think that the law has the weight of one of the major physical laws like conservation of momentum, or mass or electric charge. To do so, however, is to make a very serious mistake. There are numerous exceptions to the law in the conduction of liquids, gases and non-metals. Examples of non-linearity are even more common and arise often because of the heating effect of a current. Household examples include the wires in electric fires and in electric lamps, where the high operating temperature of the wire changes the mobility of the valence electrons so that they drift less readily than when cold. The wire, as a result,

linear elements

Ohm's law

electrical resistance
ohm

Wire resistance is $3 \times 4 = 12 \ \Omega$ (ohms).
Voltage required is $V = IR$
$\qquad\qquad\qquad = 2 \times 12 = 24$ V.

$R = \dfrac{V}{I} = \dfrac{6}{0 \cdot 5} = 12 \ \Omega$

passes less current at its operating voltage than might be expected from its low voltage measurements (Figure 26).

Among the exceptions we have already considered are gaseous conduction, which tends to have an *I*, *V* curve like Figure 27 and, of course, the diode (Figure 28), whose current varies with both filament temperature and with the direction of the applied voltage. For all these exceptional cases we can either say that the resistance, calculated as the ratio *V*/*I*, varies with current and tabulate this or we can just tabulate the *I*, *V* relationship itself.

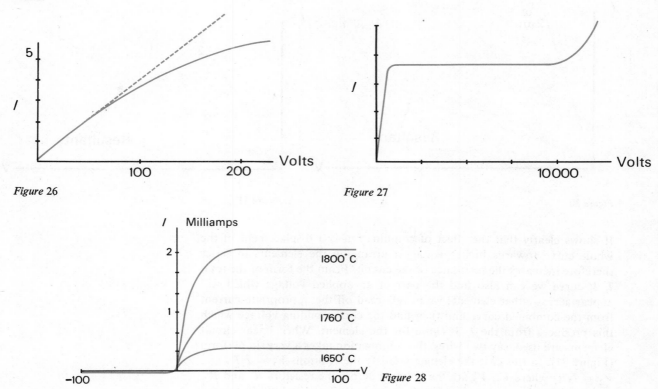

Figure 26

Figure 27

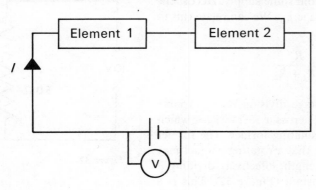

Figure 28

4.2.2 Electrical resistance of elements in combination

With a single resistor it is possible to do little more than state or calculate the current which will flow when a given voltage is applied. When resistors are compounded in networks, rather more questions may be asked. We may compute the effective resistance of the network between given points or find the voltage across some given resistor for a different set of input points. Our next task will be to find the rules which describe the flow of current in circuits comprising numbers of elements whose *I*, *V* curves are known.

Suppose first of all that we have a network of just two elements connected end to end as in Figure 29, an arrangement we call *series connection*. The

series connection

Figure 29

I, V curves for the elements, which might be a long wire and an electric motor are shown in Figure 30. To find the effective resistance of the combination, we remember first that the electric current moves in a loop: the same current must therefore flow through each element. For each level of current we can plot the total voltage across the combination by looking up the graphs to find the individual voltages appropriate to this current and adding them. The I, V curve for the combination obtained in this way is shown dotted.

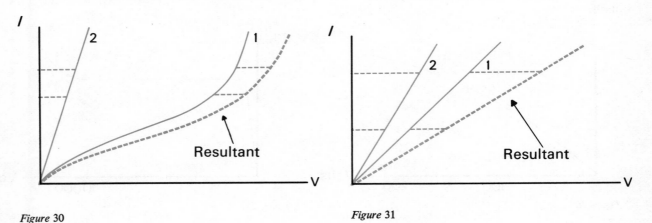

Figure 30   Figure 31

It shows clearly that the effect of combination is a displacement of the whole curve towards higher voltages: arranging the elements in series therefore increases the resistance of the circuit. From the form of the total I, V curve we can also find the part of an applied voltage which will appear across either element: we merely read off the appropriate current from the combined curve and then find the corresponding voltage which this produces from the I, V curve for the element. When linear circuit elements are used, say two wires, the combination takes a very simple form (Figure 31). In this case the elements satisfy the relations $V_1 = I_1 R_1$ and $V_2 = I_2 R_2$ where V_1, V_2 are the voltages across the resistors R_1 and R_2, and I_1, I_2 are the currents passing through them. Since we have just noted that in series combination the circuit elements must pass the same current I and that their individual potentials are summed to give the total potential V we must have:

$$V = V_1 + V_2 = IR_1 + IR_2 = I(R_1 + R_2) = IR,$$

where $R = V/I$ is, by definition, the resistance of the combination. The combination law for resistance is therefore simply $R = R_1 + R_2$. For this case we find the resistance of the total by summing the resistances of the individual elements. Evidently the same form of relationship will hold if many elements are put in series.

The series circuit is often used to divide voltages, that is, to provide a voltage which is a fraction of that available from some supply. Across the resistance R_1 of our calculation there is a voltage V_1. We can relate this to the applied voltage V by:

$$V_1 = IR_1 = \frac{V}{R} R_1 = V \frac{R_1}{R_1 + R_2}$$

Your home kit uses exactly this idea of voltage division in its variable power supply. A constant 10 volts is provided across a 50Ω resistor which takes the form of a wire wound on to an insulating former. The former has been bent into a circle and is arranged so that a rotating contact may touch the wire at different points along its length, effectively dividing it into two variable resistors whose sum is constant (Figure 32). This is a

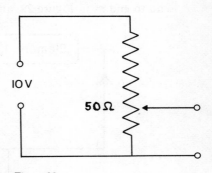

Figure 32

32

device we call a *potentiometer*. The voltage between either end of the wire resistor and the rotating contact may thus vary over a 10-volt range. We have arranged the internal connections so that the actual range provided is from $+5$ volts to -5 volts: in effect the zero point of your supply is connected to the mid-point of the resistor.

potentiometer

Exercise

You have a 100 Ω potentiometer, a 2-volt supply and a range of resistors. How can you construct a voltage supply which is variable over the range 0–0·01 volts?

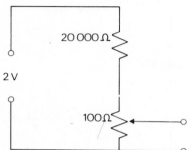

Figure 33

Put a resistor in series with the potentiometer so that 0·01 volts are developed across 100 Ω for an input voltage of 2V (Figure 33). Using the equation above, we need to find R_2 so that $V_1 = 0\cdot01$ V when $V = 2$ and $R_1 = 100$ Ω:

$$0\cdot01 = 2 \times \frac{100}{100 + R_2}$$

Thus $R_2 + 100 = 20\,000$ Ω and $R_2 = 19\,900$ Ω.

Effectively we can take $R_2 = 20\,000$ Ω.

Another possible arrangement of circuit elements is the *parallel connection* shown in Figure 34. With this arrangement the current I supplied by the battery is split into two currents, I_1 passing through element 1, and I_2 through element 2. Since charge cannot accumulate within the circuit, we know that:

$$I = I_1 + I_2,$$

parallel connection

and since the ends of the elements are connected together both must experience the same potential difference V.

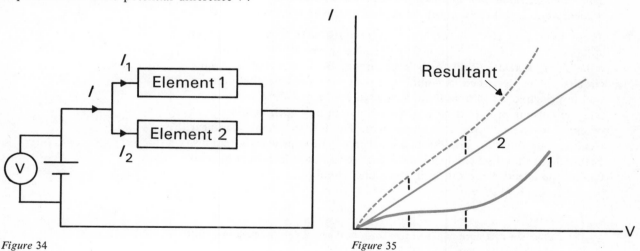

Figure 34

Figure 35

Combination of the elements 1 and 2 for this case is accomplished as in Figure 35. For any given voltage we add the currents through the individual elements to find the total current and in this way build up the resultant I, V curve. This time the displacement of the curve is towards the lower voltages, indicating that the effective resistance is reduced. As before, the part played by the individual elements can be distinguished: for a given total current the combined I, V curve gives the appropriate voltage, and reference to the individual curve gives the corresponding current in the elements. For linear elements this form of circuit also offers a simple solution for the total resistance. We perform the graphical construction using the known I_1, V_1 and I_2, V_2 relationships. This time we write the Ohm's law relationship in the form $I = V/R$ and put:

$$I = I_1 + I_2 = \frac{V_1}{R_1} + \frac{V_2}{R_2} = \frac{V}{R_1} + \frac{V}{R_2} = V\left(\frac{1}{R} + \frac{1}{R_2}\right) = \frac{V}{R}$$

where R is again the effective resistance of the combination defined by the relation V/I. Thus,

$$\frac{1}{R_1} + \frac{1}{R_2} = \frac{1}{R} \text{ or } R = \frac{R_1 \, R_2.}{R_1 + R_2}$$

This is the formula for the combined resistance. Note that in this case the resistance of the combination is always smaller than either of the individual resistors, whereas in the series case it was always greater.

Exercises

A 5 Ω and a 3 Ω resistor are connected in parallel. What is their resistance in combination?

$$\frac{1}{R} = \frac{1}{R_1} + \frac{1}{R_2} = \frac{1}{3} + \frac{1}{5} = \frac{8}{15}$$
$$\therefore R = \frac{15}{8} = 1\tfrac{7}{8} \ \Omega$$

A 1 000 Ω and a 10 Ω resistor are connected in parallel. What is their resistance in combination?

$$\frac{1}{R} = \frac{1}{R_1} + \frac{1}{R_2} = \frac{1}{1\,000} + \frac{1}{10}$$
$$\therefore R \approx 10 \ \Omega$$

With resistors connected in parallel, the overall resistance tends towards the resistance of the smallest element. *Note the contrast here; a series combination with a particularly* large *resistor will have a resistance approximated by the value of the large resistance: a parallel combination with a particularly* small *resistor will be approximated by the small resistance.*

These series and parallel combinations of resistances may seem rather specialized and trivial, and not to reflect the complexity needed to describe the I, V relationship for a large circuit. But this is not so. Many circuits can be described as combinations of series and parallel units so that in principle the technique we have used above will allow us to calculate the current produced by a given voltage in any such combinations of circuit elements—even non-linear ones. The task is simple for linear elements, although for non-linear ones it might be depressingly tedious. Figure 36(a) shows schematically how the combination laws may be used to find the effective resistance of a complex circuit.

Exercise

Find the resistance of the array shown in Figure 36(a) assuming that each individual resistance has a value of 10 Ω.

For answer, see Figure 36(b) on page 36.

4.3 Power and heat in a circuit

The most obvious of all effects associated with the passage of an electric current must be the production of heat. We are familiar with it domestically in electric fires, lights and fusewires, and less obviously in the warmth of an electric motor or a television set. To produce this heat, or to produce the mechanical power of a motor, electrical energy must be used in the circuit. Our purpose in this section will be to relate this release of energy

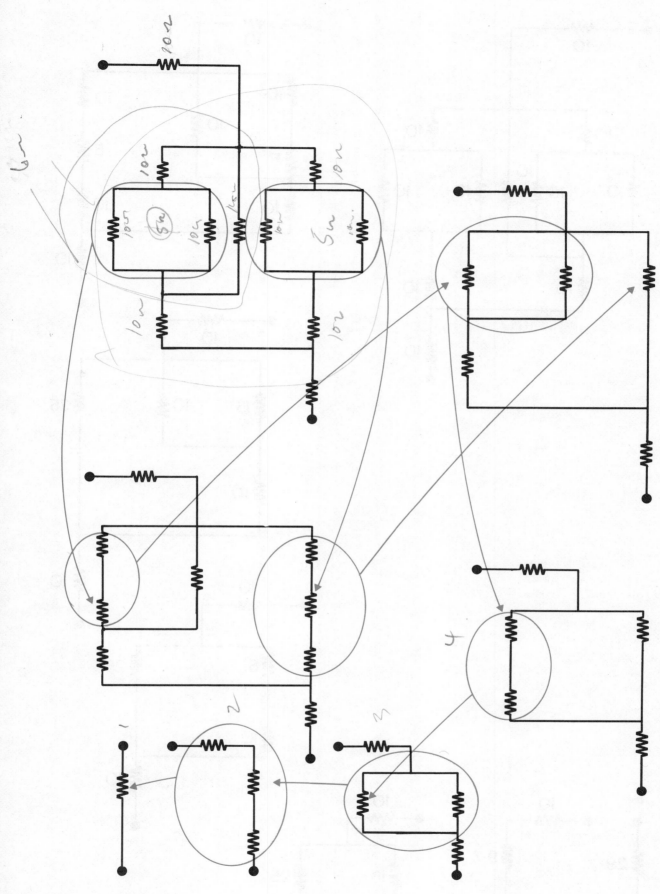

Figure 36(a)

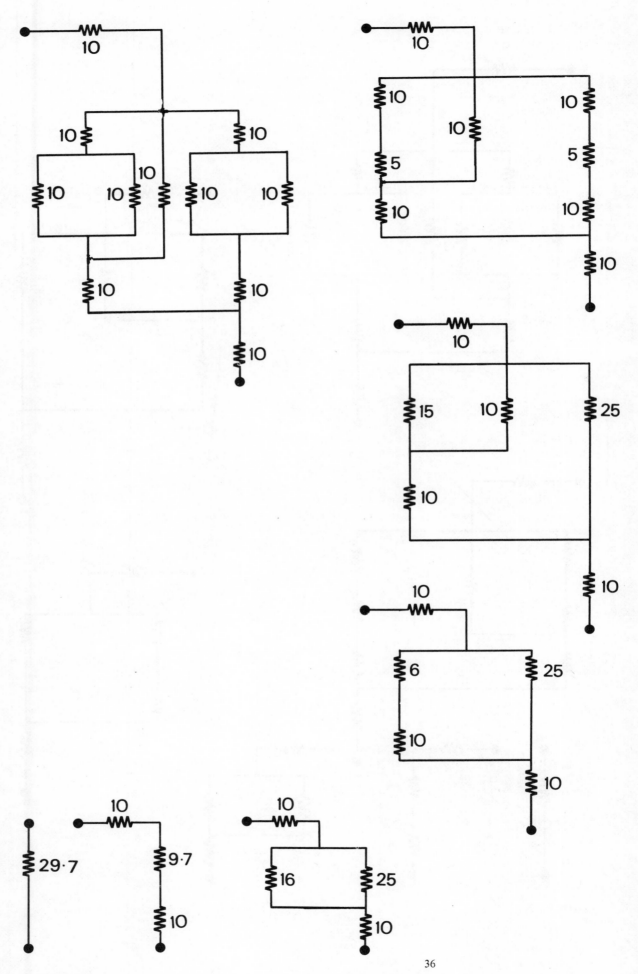

36

Figure 36(*b*)

to the electrical characteristics of a circuit: to the current, voltage or resistance of the elements involved.

In Section 2.3.2 we laid the basis of this calculation when discussing the flow of charge through a metallic conductor. An electron moving from a position of high potential energy to one of lower potential energy was supposed to bump its way through the material, steadily converting its potential energy into heat as it did so. We can put this idea on a quantitative basis by returning to our definition of electric potential. The potential difference between two points in a circuit was defined as the energy which would be required to move one coulomb of charge from the low voltage point to the high voltage one. If a charge of Q coulombs is allowed to flow naturally through a conductor it will do so from a high voltage point to a low one. If in so doing it passes through a potential difference of V volts, QV joules of energy will be released.

Often it is the rate at which this energy is released that is the most important consideration. We can calculate this by working in terms of current rather than charge. Suppose that the charge Q passes at a steady rate during a time of t. s. This will correspond to a steady current $I = Q/t$ A. The energy released per unit of time is therefore:

$$P = QV/t = IV \text{ joule sec}^{-1}$$

In both mechanical and electrical usage the rate at which a device uses energy is called its *power*. The rate of use which corresponds to one joule per second, we call the *watt*. We shall use this as our unit of power from now on.

power

watt

The calculation we have just completed shows that electrical power is given by the product of a current with a potential difference. This is a general result and allows us to calculate the power in any circuit element regardless of the type of element and the details of how the power is used. It does not matter that the specific process involved produces mechanical motion, accelerates electrons through a vacuum or heats a conducting wire: the power used is always given by the product of current and voltage.

Often it is convenient to express power in terms of a current or a voltage only. To do this we need only remember that $V = IR$ and write:

$$P = IV = I \times IR = I^2 R$$

$$\text{or} \quad P = IV = \frac{V}{R} \times V = \frac{V^2}{R}$$

The expressions are equivalent and are used, as convenient, for any specific problem.

Exercise

A power station produces 2 000 MW (mega watts, i.e. 10^6 watts) and delivers it to a distant town at a potential of 400 000 V. What current passes and, as seen from the power station, what is the apparent electrical resistance of the town?

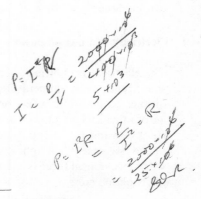

Power: $P = VI$

$$\therefore I = \frac{P}{V} = \frac{2\,000 \times 10^6}{400\,000} = 5\,000 \text{ A}$$

Resistance: $P = \frac{V^2}{R}$

$$\therefore R = \frac{V^2}{P} = \frac{(4 \times 10^5)^2}{2 \times 10^9} = 80 \ \Omega$$

The range of electrical power that we normally encounter outside industry (Figure 37 (a)–(d)), is about 0·5 W up to 10 000 W (10 kW). Torch bulbs are among the least powerful things we use; typically they run at about 0·6 W, corresponding to a current of 0·2 A from a 3 V battery. Room lights, at about 100 W from a 240 V supply, rather surprisingly, use less current— about 0·4 A—despite the much greater power. Refrigerators, vacuum cleaners and hair driers use progressively greater amounts of power until we come to the electric fire, which commonly uses 1–3 kW, and the electric cooker, which may use 10 kW.

An interesting and non-electrical comparison on the domestic scale is the motor car, which is typically advertised as producing about 50 horse-power. As one horsepower is equivalent to 746 W in electrical terms, cars run at an electrical equivalent of about 40 kW. Even this figure is an under-estimate, for much more energy is dissipated by a motor as heat than is produced as mechanical energy (this point will be enlarged upon in Unit 20, Energy conversion) so cars are extremely heavy users of power, at least by domestic standards.

When you purchase electricity, or rather when you pay for the electrical supply connected to your home, the basis for the accounting is a standing charge, which in principle helps to cover the cost of the equipment provid-ing the supply, and a charge proportional to the amount of energy which you have received. In terms of the calculations we have been doing we might expect this energy to be expressed in joules, or even possibly in coulombs since the supply voltage is constant at 240 and we therefore obtain 240 joules from each coulomb which passes.

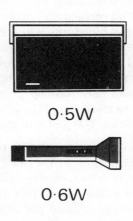

0·5W

0·6W

Figure 37(*a*)

But not everyone is a scientist, prepared to work with units rather distant from daily life, so a more homely unit is used in preparing our bills. It is called the *kilowatt-hour* (kWh). The idea of this unit is that many domestic appliances use power at a rate close to one kilowatt: the kilo-watt-hour is the energy used by such an appliance running at a rate of a kilowatt for an hour. Since a joule is the amount of energy used in running at a power of a watt for one second, the kilowatt hour is $1\,000 \times 60 \times 60 = 3\cdot6 \times 10^6$ J.

kilowatt-hour

For many purposes the kilowatt-hour is a good unit for putting a mental scale to a process. It is the energy used by a small single-bar electric fire in one hour; it is also quite close (1 000 versus 746) to that traditional idea of energy introduced by James Watt, the amount of mechanical work which can be done by a horse in one hour.

4.4 Electricity and measurement

The electrical properties, such as resistance, of most materials vary quite significantly with temperature, with pressure and with other physical con-ditions. In addition, a few of the materials also develop small potentials across them when subject to radiation or stress; examples of this are the solar cell and the sound meter of your home kit which produce voltage, in the one case when light shines on a prepared surface, and in the other when mechanical force is exerted on a small crystal by the vibrations of the microphone cone.

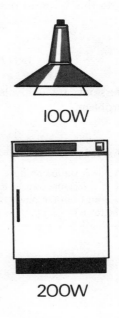

100W

200W

Figure 37(*b*)

There are many effects of this general character. For purposes of measure-ment, what matters is that some of them can be reproducible. A platinum wire with a resistance of 100 Ω at 20°C has a resistance of 137 Ω at 120°C and 174 Ω at 220°C; its resistance changes by about 0·37% for a 1°C rise in temperature. Similarly, a manganin wire (an alloy of copper, nickel and managnese) increases its resistance by about $2\,10^{-7}\%$ per atmosphere of pressure (about 10^5 Pa)—not very much perhaps but enough to show at high hydraulic pressures. Because these variations are found to be re-producible, the effect is used inversely, the argument being that when the platinum wire has a resistance of 137 Ω it must be at a temperature of 120°C, and so on; the electrical effect is taken to be the measure of the physical phenomenon.

Measurements of this sort never have justification in absolute terms; their calibration does not derive, for example, from an elaborate calculation in solid state physics of the temperature variation of the electrical resistance of platinum. Such theories are generally incapable of reaching the precision with which the comparative measurements can be reproduced. The measurements are therefore calibrated relative to a number of set points whose position on the appropriate scale of temperature, pressure, light or sound level, has been determined in independent experiments.

500W

Figure 37(c)

Because electrical measurements can be made easily, and with both sensitivity and precision, this process of calibrating apparently obscure electrical effects as measures of non-electrical phenomena has become very common. A multiplicity of devices which change mechanical displacement, temperature, pressure, sound and light levels, liquid levels, humidity, weight and speed into electrical outputs is now available. We call such devices, which have this ability to change effects from one physical form into another, *transducers*.

transducers

This unit is not the place to elaborate upon the vast and fascinating array of transducers currently in use. The development of new variations is a task which seems to appeal to the inventive mind—would you have thought to use an expanding human hair to indicate humidity—and it would be a formidable task even to list those of essentially electrical character. We shall instead concentrate on just one kind of transducer, the kind which depends on changes in electrical resistance, and use it to illustrate one way in which the ideas on circuits, introduced earlier in this section, may be used to give measurements of considerable sensitivity with apparatus of a quite unsophisticated nature.

1 kW

To detect temperature changes using a platinum resistance thermometer requires a sensitivity to resistance changes of about 0·3%, if changes of 1°C are to be detected. Other materials may be less demanding; iron would only require 0·5% sensitivity while the thermistor in your home kit may require just 1–2% sensitivity. Whatever the material, however, quite small resistance changes must be detected if effective temperature measurements are to be made. To make useful measurements of pressure or mechanical strain, comparable, or even finer, measurements are normally required.

2–3 kW

Something of the difficulties involved in making this sort of measurement emerges if we work out some of the numbers likely to be involved. To measure a resistance, we need to measure a current and a voltage. At a sensitivity of better than 1% for each reading this is not easy, so we simplify a little by adding a resistor in series with the transducer (Figure 38) and comparing the voltage across each. Since they pass the same current, the ratio of the voltages is also the ratio of the resistances.

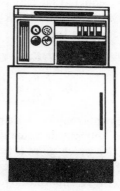

8–10 kW

Imagine that we set out to do the experiment in this way. We clearly get more volts across the transducer by using a large current and by keeping the other resistance small. Neither of these choices, however, optimises the measurement process. Any current will generate heat, warming the transducer, while reducing the value of the comparison resistor just lowers the precision of the second voltage measurement. The best choice seems to be to have the maximum current which does not significantly heat the resistor and to have two resistors of comparable value. Now we know that an electric fire uses about 1 kW, a light 100 W and a torch about 0·5 W so a first guess at a safe power level might be $\frac{1}{10}$ W. With resistors of 100 Ω this means $I^2R \approx \frac{1}{10}$, $I \approx \frac{1}{30}$ A and there is almost 3 V across each resistor.

50 kW

Figure 37(d).

In an actual experiment the acceptable power level, which depends on the size of the resistor and the degree to which it is thermally insulated from its surroundings, can only be found by measuring the transducer resistance for a series of currents. The effect of heating will then show as a resistance increase with current.

In our small circuit it now seems that a supply of 6 V, with roughly 3 V across each resistor, is about the optimum. Even now, however, we do not have a very acceptable measuring system. If we had voltmeters reading to a maximum of just over 3 V with scales of 100 divisions, a 1°C temperature change would produce a variation of only $\frac{1}{3}$ division, or 1 millivolt, with a platinum resistor; with a manganin pressure gauge a pressure change to twice atmospheric would not visibly move the pointer.

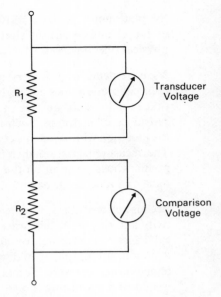

Figure 38

What is needed is a way of discarding the 3 V across each resistor and of displaying only the *changes* in voltage as the resistance of the transducer is modified by pressure or temperature. To do this we must mark the potential of the point between the resistors when the transducer is in some standard condition—ice temperature or atmospheric pressure, etc.— and compare it with the value it actually takes under other conditions. A second pair of resistors connected across the voltage supply (Figure 39) provides the means of doing this. With this circuit there must always be some choice of resistors with which the meter will record no current. This establishes the condition of zero potential difference between the junction points of the resistor pairs.

Subsequent variation of transducer resistance modifies the potential of the junction point on only one side of the circuit and generates a potential difference across the meter. With this arrangement there is no need to use a meter capable of reading 3 V; instead a sensitivity appropriate to the effect being measured is used. In the example we have considered a meter with full scale deflection for 50–100 mV would work very well for displaying the temperature of hot water.

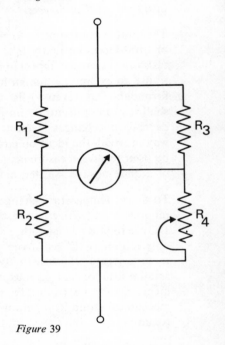

Figure 39

As described, our measuring circuit indicates by a small current passing through a meter. In this form the circuit is often used for control purposes. For measurement, a slight elaboration is preferred. Whatever value the transducer resistance may take, some value of the variable resistance may be found which brings the meter to a zero reading. At this point the junction points of the resistance pairs exactly divide the supply voltage in the same ratio: the voltage across R_1 just equals that across R_3 and that across R_2 equals that across R_4. Now the current in R_1 and R_2 is the same, say I_1 while that in R_3 and R_4 is also the same, say I_2. When the meter reads zero then:

$$I_1 R_1 = I_2 R_3 \quad \text{and} \quad I_1 R_2 = I_2 R_4$$

$$\text{so} \quad \frac{I_1 R_1}{I_1 R_2} = \frac{I_2 R_3}{I_2 R_4} \quad \text{or} \quad \frac{R_1}{R_2} = \frac{R_3}{R_4}$$

Measurements may be made in terms of the resistors only. With R_1 and R_3 fixed and known, the value of R_2, the transducer, can be found from the value of R_4 which gives a zero reading. We call this condition, the condition of *balance* in the circuit and we call this configuration of resistance and meter a *resistance bridge*.

resistance bridge

This form of circuit has the great advantage that the balance point is independent of fluctuations in the supply voltage. In addition, only one reading is needed to determine the balance condition as opposed to the two needed to measure a resistance. In its many variations, the bridge circuit must surely be the most common in the whole of measurement technology.

Section 5

Magnetic fields and electromagnetic forces

So far we have made only passing reference to magnetism. You may wonder why it was classed with electricity in the title of these units, when clearly magnets themselves are of trivial industrial importance beside electricity. The function of this section is to redress this state of affairs; to show that the study of magnetism is limited and narrow in the absence of a knowledge of electricity, the study of electricity being incomplete without reference to the magnetic effects which it causes.

5.1 Permanent magnets

We begin with magnets. The most familiar of magnets must be the compass needle. Its prime property is a tendency—not very strong—to rotate about its vertical axis until it is aligned almost along a north–south direction. If a compass is brought near another magnet its orientation will change. Even the small magnets found in toys will align a compass in a way which is clearly more forcible than the north–south alignment brought about by the Earth.

We can describe the magnetic effects of a heavy magnet by drawing maps showing compass orientation in the region surrounding it. Figure 17 shows this for a bar magnet. The lines of the map have been drawn so that the tangent at each point is along the line that would be taken by a compass at that point. The lines therefore indicate the paths which would be traced out by a compass moved progressively in the direction indicated by its north-pointing tip.

Maps of this sort vary in form with the shape of the magnet used (Figure 40(a) and (b)), but they have some properties in common. With any magnet there will be a region from which the north-pointing tip of a compass will always point away and towards the south-pointing tip will be attracted. Somewhere else on the magnet there will be a region where the opposite occurs. The lines traced out by our maps join these regions.

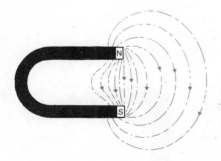

Figure 40(*a*)

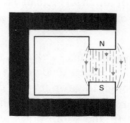

Figure 40(*b*)

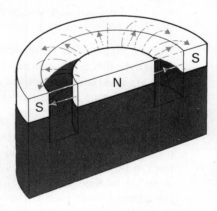

Figure 41

Not all magnets have a north–south seeking tendency (the one in Figure 41, for example, has none), but a freely suspended bar magnet will always align north–south. If we mark the north-seeking end, or north pole, of a magnet and then draw an orientation map around it we will always find that the end which repels the north pole of the compass is the north pole of the magnet and the end which attracts it is the south pole. The maps therefore confirm another well-known property of magnets: that in certain positions a pair of magnets will attract while in others they will repel (Figure 42 (a) and (b)). We now require that they attract when unlike poles are placed together and repel when like poles are placed together.

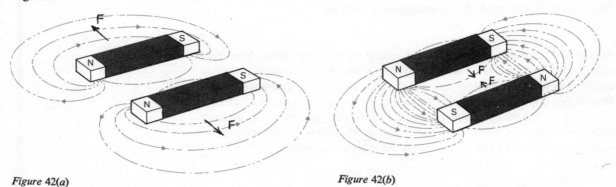

Figure 42(*a*)   *Figure* 42(*b*)

A law which says that like things repel and unlike attract, presents an almost irresistible temptation to argue by analogy. We shall not give in to it: we shall not introduce the idea of a magnetic charge, because it is a concept which has proved historically not only to be unfruitful but also, as we shall see, unnecessary. We shall permit analogy only in the name we give to the region of magnetic influence surrounding a magnet, which we shall call a region of *magnetic field*.

magnetic field

When we dealt with electrical forces we used the idea of an electric field. This was the force which would be exerted on one unit of electric charge placed at some point in the region of electric field. We shall use a similar idea for magnetic field, but shall centre our ideas more strongly on the orientation maps we have been using. As with the electric field, we have a quantity which varies both in direction and in strength. We shall regard the field direction as the direction taken by the north pole of a compass; its strength we shall take as a measure of the tendency of the compass to return to its initial position if displaced. In regions where the lines of the map lie close together, which we shall describe as regions of high *magnetic flux density*, this tendency is strong, and where they are widely spaced (low flux density) it is weak. We can therefore regard the maps as giving a full indication of magnetic field, the direction being shown by the direction of the lines and the strength being proportional to their density. We shall come to a more quantative expression for the strength later.

magnetic flux density

The last major effect we associate with magnets is their attraction for iron. Unlike the forces between pairs of magnets, the forces between magnets and iron are invariably attractive. This is not an effect we can explain without recourse to our atomic model once more, but we can give a description of it in terms of field maps.

When iron is placed in a region of magnetic field it has a strong effect on the overall pattern of field lines. The general tendency is for the field lines to be shifted so that they pass through the iron leaving other regions denuded of field (Figure 43). If the iron is arranged so that it almost completes a wide path between the north and south poles of the magnet the region of observable magnetic field is almost completely concentrated in the gap, and large forces act between the magnet and the iron. When contact is made, with the iron bridging the magnet poles, the force needed for separation is still stronger.

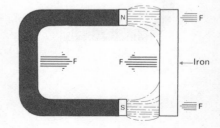

Figure 43

In the magnets used to hold workpieces to the bed of a milling machine, or to secure the door of a cupboard or refrigerator, advantage is taken of these large forces, and large flat pole surfaces are provided against which a flat piece of iron may make good contact (Figure 44). One rather interesting feature of these large attractive forces is that they are substantially reduced if another magnetic path is provided. In Figure 45, contact of the heavy piece of iron with the magnet will effectively release the finer piece at the other side. Such ideas can be of use in removing delicate parts from a magnet, since a permanent magnet cannot, after all, be switched off.

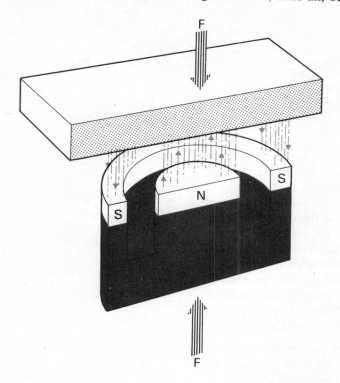

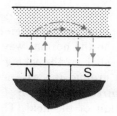

Figure 44

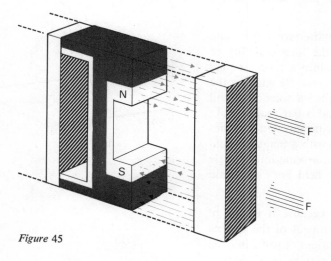

Figure 45

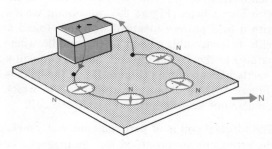

Figure 46

5.2 Electromagnetism

A remarkable unification of electric and magnetic effects is seen if a compass is placed close to a wire carrying a current. Figure 46 shows the orientations that might be seen in a set of compasses laid on and under a wire carrying a substantial current—say 10–20 A. The compasses completely lose their north–south orientation and take up directions determined only by the orientation of wire and their position relative to it. The current flowing in the wire has produced a magnetic field in the region surrounding it.

43

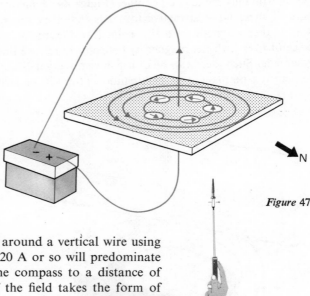

Figure 47

The field pattern is best seen if it is mapped around a vertical wire using compasses in the normal way. A current of 20 A or so will predominate over the usual north-seeking tendency of the compass to a distance of 10–20 cm around the wire and the map of the field takes the form of circles centred on the wire (Figure 47). Reversal of the current, reverses the direction of the field. The field direction always circles the wire in the direction which would be traversed by one's hand in driving a right-hand screw along the direction of the current. This form of map is quite different from anything ever found with permanent magnets because the field lines form closed paths around the wire. With the permanent magnets the lines always go from south pole to north pole or else disappear in a north–south direction following the magnetic field of the Earth.

More detailed experiments show that the effects depend truly on current. Wires of differing electrical resistance produce fields reaching out to the same distance, and even a tube of electrolyte, using both positive and negative ions instead of electrons for its conduction, gives similar effects for similar currents.

Because wires are readily twisted, it is a simple matter to make quite complicated field patterns. Figure 48 shows how the magnetic flux of Figure 47 may be concentrated by bending a wire into a loop. It may then be intensified further by adding turns to make a coil. By winding the wire into a helix (Figure 49), we obtain what is called a solenoidal coil with a field pattern identical with that found around a bar magnet, but with an observable region in the middle where a concentrated flux lies parallel to the axis of the coil. Such a coil will align with a magnetic field exactly like a compass (Figure 18). If the helix is bent around into a circle (Figure 50), we obtain a toroidal coil in which the field lines are again circular but lie concentrated within the coil.

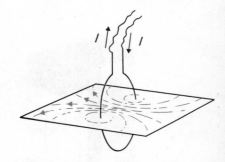

Figure 48

The toroidal coil is of particular interest nowadays because it is one of the arrangements proposed for the magnetic containment of the plasma involved in nuclear fusion reactions. This is also the origin of some interest in the double coil illustrated (Figure 51), which in plasma technology is

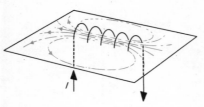

Figure 49

Figure 50

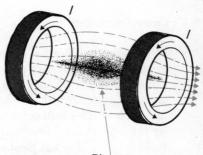

Figure 51   Plasma

called a magnetic mirror. It has the property of tending to contain plasma in the region between the coils.

Iron modifies the magnetic field around a current-carrying coil just as it does that around a permanent magnet. The main difference is that, being hollow, the coils allow the iron to be put right into the core and to form complete loops, where with a permanent magnet at least part of the path

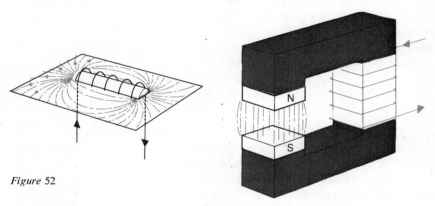

Figure 52

Figure 53

must be of the magnet material. If iron is added to a hollow solenoid there is a striking increase in the flux density of the surrounding region for a given coil current (Figure 52). If the iron is bent around almost into a complete loop the flux is further increased and is concentrated in the gap (Figure 53). Devices of this type in which a large flux density is produced by the current passing through a coil wound on an iron former are called *electromagnets*. As with the iron bridge on the permanent magnet a large force is exerted between the exposed faces. This is the origin of the force which drives magnetic actuators, for example the Post Office relay (Figure 54).

electromagnets

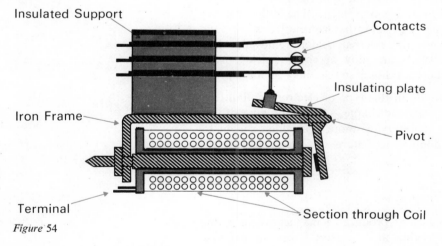

Insulated Support

Contacts

Insulating plate

Iron Frame

Pivot

Terminal

Section through Coil

Figure 54

5.2.1 Magnetism on an atomic scale

The magnetic properties of a current loop suggest an origin for permanent magnetism which we can relate to our atomic model. The path of a moving electron is effectively the path of an electric current, so an electron moving in a closed path may fairly be regarded as having all the properties of a current loop. Among these should be included the magnetic properties of a loop which would require that the electron produce a field in the region around it similar to that of Figure 48.

45

Since the atomic model requires most of the electrons in an atom to orbit a positively charged nucleus we clearly have the possibility, by aggregation, of producing substantial magnetic effects if only electrons will orbit in parallel loops like the coils of Figure 49. While magnetic effects do arise in this way, it turns out that they are never large because, in any normal piece of material, as many electrons orbit in one direction as in the other, and there is a high degree of cancellation.

While small effects due to orbital motion of electrons can be identified, they are much weaker than the effect we commonly refer to as magnetism. This stronger effect, more properly called *ferromagnetism*, also has its origins in electron motion, but on a smaller scale. Electrons do not have a structure that we can recognize or describe, but they do have more properties than just charge and mass. The property of importance to magnetism is called spin.

ferromagnetism

All electrons have inertial properties, in addition to any orbital motion, which allow us to treat them like spinning tops of charged material. On a minute scale therefore, individual electrons may, in themselves, be treated as current loops and if many electrons can be aligned with their spins parallel we may expect large-scale magnetic effects to show. It is alignment of this type that takes place in iron, nickel, cobalt—the materials we know as being magnetic. What is puzzling is that the ferromagnetic materials show this effect, while others, like copper or zinc do not. The answer lies in some details of the atomic structure. Most electrons tend to be paired in their orbits with their spins opposed, and the net effect is an annulment such as would be found if coils carried equal and opposite currents. In the magnetic materials some of the electrons are unpaired, iron atoms for example having four unpaired electrons.

Gross magnetic effects arise because the spins of the unpaired electrons may be aligned parallel with each other by the action of an external magnetic field. The field aligns a few of the electron spins and these then augment the overall field, turning more and more of the spins into alignment with them in a rapid co-operative movement. The difference between iron and permanent magnet material is that iron electrons lose their common orientation readily when the external field is removed, while permanent magnet material (usually an alloy of cobalt, nickel, iron, copper and aluminium called Alnico) is chosen so that this will not happen.

It is because magnetic effects reside in this type of electron motion that we did not develop the idea of magnetic charge, and it is because of this that the many experiments of a century or more ago which sought to isolate magnetic charge were doomed to failure.

5.3 Electromagnetic forces

As a wire carrying an electric current produces a magnetic field in the region which surrounds it and is thus able to exert forces on magnetic material in the region, it is reasonable to guess that magnetic forces may also act on the wire.

One circumstance in which a force may be observed due to the interaction of a current with a magnetic flux from a magnet is shown in Figure 55. A magnet of the type shown has a particularly well-defined magnetic field with a region of high flux density between the pole pieces and essentially zero flux density elsewhere. The field direction is directly from the north pole-face to the south pole-face.

If a wire is placed between the pole pieces and arranged perpendicular to the field, a current in it will produce a force perpendicular both to the field and to the current. As with the field around a current-carrying wire, the direction of the force in this case is quite specific: the direction of the force on the wire is found by rotating one quarter turn in a right-hand screwing motion along the current direction from the field direction to the force direction. Obviously an equal and opposite force must act on the magnet.

(The traditional mnemonic for these directions is Fleming's left-hand motor rule in which, setting the thumb, forefinger and second finger mutually perpendicular, the *f*orefinger indicates the *f*lux, the thu*m*b the *m*otion, and the se*c*ond finger the *c*urrent (Figure 55).)

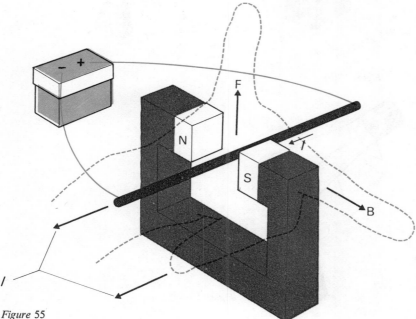

Figure 55

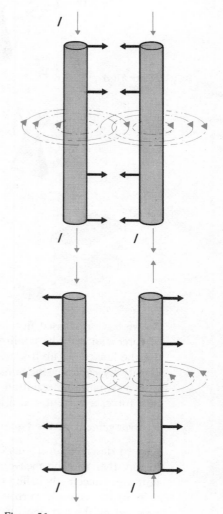

Figure 56

Only fields perpendicular to the current produce force. If a field lies in a direction skew to the wire, only the component of field perpendicular to the wire acts in producing force.

One interesting situation in which forces of this type are produced is when currents flow in wires lying parallel to each other (Figure 56). As we already know, around each wire there is a region of magnetic field, the field lines circling the wires. Each wire thus lies perpendicular to the field produced by the other—exactly the arrangement which we have agreed will produce a force. Applying the left-hand rule or the field–to–force corkscrew rule, we find that the forces must be of attraction when the current directions are the same and of repulsion when they are opposite. As might be expected, the forces are always equal and opposite.

Although forces of this type are strong enough to be measured quite easily in laboratory conditions, they are too weak to be obvious in normal electrical circuits (Figure 57(a) and (b)). They become important mainly when very high currents are used, for example in the windings of heavy electromagnets or between the wires of transmission lines when they are momentarily subjected to overload conditions. Figure 58 shows one small-scale example of forces between currents which is of especial interest because it confirms that we may fairly treat a stream of electrons as being equivalent to a flow of electric current in the opposite direction.

It can be shown experimentally, and it is easy to appreciate, that the force on a wire is proportional to the current passing, to the length of the wire lying within the magnetic field and to the density of the magnetic flux. If we call the current I, the length of wire between the pole faces l and the flux density B, the force F on a suitably arranged system of wire, current and field is: $F = \text{const.} \times B\,I\,l$.

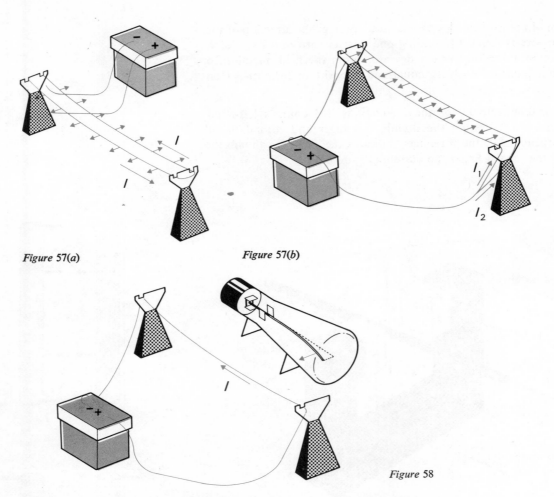

Figure 57(a)

Figure 57(b)

Figure 58

As we have discussed flux density only in terms of comparison, this law of force can only be confirmed as a proportionality. We can, however, use the law to set up a scale of *flux density*. We can *define* a unit of flux density by *choosing* the constant in our equation to be 1. This leaves us with a remarkable equation in which a flux density is described in terms of a force, a current and a length: $F = B I l$.

flux density

Rearrangement of the equation to give: $B = \dfrac{F}{Il}$ shows that flux density may be described in terms of a *force* divided by a *current* and a *length*: it may therefore be measured in units of newton ampere^{-1} metre^{-1}, a unit more succinctly called the *tesla*. It is the implication of this relationship which is most surprising: it states that our basic description of magnetism, the flux density, is fixed by the units of mass, length, time and electric charge. Magnetic effects may be completely described in terms of the units of electricity.

tesla

5.3.1 Forces on a coil

Because so much of our industrial machinery depends for its smooth operation on our skill in making good rotating bearings, the technology for producing rotary motion from electromagnetic forces developed long before that for producing linear motion. As a result, some of the most important uses of electromagnetic force involve the effect of a magnetic field on a coil carrying a current. When a current passes through a rectangular coil, threaded as shown in Figure 59 by a magnetic flux of uniform density B, forces are exerted on the whole length of the coil. Those acting on the horizontal parts of the wire are equal in magnitude, opposite in direction, and directly opposed; they therefore cancel. Those acting on

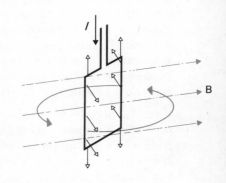

Figure 59

the vertical parts are equal in magnitude and opposite in direction but they are not directly opposed and so tend to turn the coil about its vertical axis. We refer to a pair of forces acting in this way as a *couple*. Coils of many turns experience equal forces on each turn and produce proportionately greater couples.

Two devices of central importance to our use of electricity use the couple produced by a coil. They are the moving-coil ammeter and the direct current motor.

5.3.2 Moving-coil ammeter

The idea underlying a moving-coil ammeter is to generate a couple by passing a current through a coil suspended in the magnetic field of a permanent magnet, and then to measure this couple by causing it to displace a fine spring. The essentials of the meter are therefore a freely suspended coil, a magnet arrangement providing constant flux density for all positions of the coil and a hair-spring providing a restraining couple proportional to the displacement of the coil from its rest position. A pointer attached to the coil allows its displacement to be calibrated directly in terms of current. A relatively robust instrument embodying just these ideas is included in your home kit and it gives a full-scale deflection for a coil current of only 100 μA.

Some constructional details of the moving coil meter are of interest because they illustrate the freedom with which magnetic fields may be tailored to meet special demands. The usual arrangement is to have a rectangular coil pivoted on bearings and restrained by hair-springs as in Figure 60. Often the current reaches the coil·through the springs, which are normally of phosphor bronze.

The most interesting feature of the construction is the magnetic field. When calibrated, we want equal intervals on the meter scale to correspond to equal increments of current. To achieve this, the turning effect (couple)

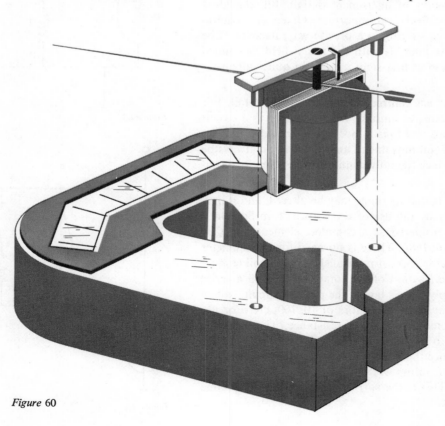

Figure 60

of the electromagnetic forces for a given current must not vary with the orientation of the coil. This demands not just that the forces remain constant as the coil turns, but also that they rotate with it. Figure 61 illustrates this: case (a) shows the forces in a constant field of fixed direction, case (b) shows the forces that we want and the local direction of the field which will achieve them. We build a field of the pattern we want by inserting a cylinder of soft iron between the shaped polepieces of a horse-shoe magnet. This gives field lines which are radial in the space, between cylinder and

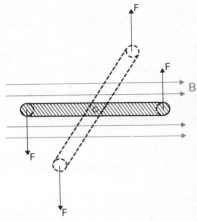

Figure 61(a)

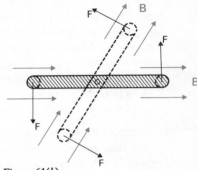

Figure 61(b)

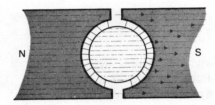

Figure 62

magnet (Figure 62) and gives forces of precisely the correct orientation and magnitude on a coil free to rotate in the space.

Another piece of field tailoring which is used in a meter concerns its overall calibration. The magnets, coils and scales are normally made in large quantities and are subject to some variation from piece to piece, yet we expect a meter to be correctly calibrated when we use it. To allow for these small variations an iron strip is connected across the magnet poles outside the coil region. By pressing this strip against the poles with a screw, a varying amount of magnetic flux may be diverted from the coil region through the strip without disturbing the overall field pattern. This gives a fine control to the field strength and allows a 1 Amp ammeter to be adjusted to read precisely 1·0 A when a calibrated current passes through it.

Now for the purposes of this course we are not at all concerned that you should learn these intricate details of instrument design. But the ideas involved in specifying magnetic fields are important. Given the requirement for fields of a precise level, varying in a precise way, means can be found to produce them even though, as with the radial field demanded around the coil, the specification at first sight may seem quite unreasonable.

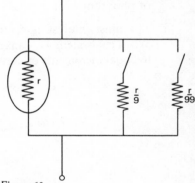

Figure 63

This discussion of the ammeter has centred on constructional detail. For most purposes, however, an ammeter must be treated simply as a circuit element just as we treated lamps and fires in Section 4. Provided that we neglect the dynamic aspects of coil movement, an ammeter can always be regarded simply as a resistor, since the only connection between its terminals is a wire coil.

A nice feature of the ammeter is that its range may be altered, and it may be used as a voltmeter, merely by the use of a resistor. From Section 4 we know that an ammeter, considered as a complete element, must lie within the main line of a circuit loop. There is no need for the ammeter coil itself to carry all this current. A resistor parallel with the coil (Figure 63) will always carry the same fraction of the total current and a series of resistors as shown —just apply Ohm's law to the parallel circuit—allow the range of a meter to be multiplied by factors of 10, 100, etc. For use as a voltmeter, the ammeter requires a series resistance. The full-scale current I of a meter is related to its full scale voltage V by $V = Ir$, where r is the coil resistance. To reduce the voltage sensitivity we need only add series resistors (Figure 64) of values $9r$, $99r$, etc. to obtain full-scale voltages of $10\ V$, $100\ V$, etc. This is the sort of switching which is done in the meter of your home kit.

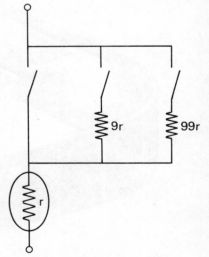

Figure 64

5.3.3 Direct-current motor

The simple coil in a magnetic field has an obvious potential for the generation of mechanical motion from an electric current. Two main problems must be resolved to make an effective motor; the first is to obtain a continuous motion, the second is to produce a reasonable amount of power in an acceptably small space.

A coil, as in Figure 59, cannot be used as a motor because it has a stable position in a magnetic field when the field direction is perpendicular to the plane of the coils. Such a coil after all is similar to a bar magnet in its magnetic behaviour and must be expected to just align itself with the field.

To obtain continuous motion, a pair of sliding contacts, usually of carbon is used (Figure 65(a)—(c)). Such an arrangement is called a *commutator*. At each half turn, the coil current is reversed so that the electromagnetic forces always act to turn the coil in the same direction. Providing the coil turns freely, and enough momentum is available to take it past the dead spots where the contact is broken, a continuous rotating motion can be achieved. To produce a motor with appreciable mechanical output, the flux density must be made as large as possible, and for this reason an iron core is again used for the coil.

commutator

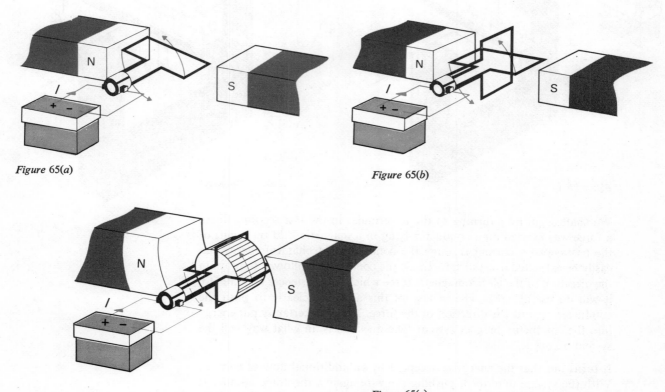

Figure 65(a)

Figure 65(b)

Figure 65(c)

Even toy motors nowadays are more sophisticated than the one sketched in Figure 65. Typically the rotating part, the *armature*, has three coils, and two magnets are used to provide the field. For all but the very smallest of motors, however, a permanent magnet is too large and heavy. Instead the magnetic field is produced by an electromagnet driven by the same power source as that which drives the armature current.

armature

In these larger motors the coil arrangement on the armature is also more complicated, the number of coils being increased to give a smoother output and more certain starting. Car starter motors are typical examples of this class of motor.

Electromagnetic induction

6.1 Induced currents

In the last section we saw that a current will produce a magnetic field in the region around it, and that a conductor carrying a current through a region of magnetic field will experience a force. These effects led us to an understanding of a number of devices, including motors and meters, whose common output was movement motivated by currents and magnetic fields. In this section we are going to look at some inverse effects in which movement and the manipulation of magnetic fields allows us to extract currents.

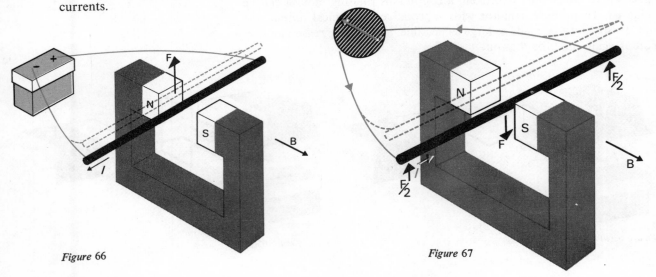

Figure 66   *Figure* 67

We shall begin by returning to the experiment in the last section, where a force was exerted on a conductor lying in a magnetic field by means of the passage of a current through the conductor. Mechanical energy can easily be extracted from such forces. If the conductor is allowed to move in the direction of the electromagnetic force which it experiences (Figure 66), it will do useful work. The inverse of this action is clearly to push the conductor against the direction of the force. This will certainly put energy into the conductor–magnet system. The question is, in what way will the system accept it?

It turns out that the energy is accepted by an additional flow of current. With the system described, pushing the wire against the force so that it moves *increases the current* above the value it has when the wire is still: this extra current therefore *increases* the *opposing electromagnetic force*. If the wire is allowed to move with the *electromagnetic force*, the *current diminishes* and so does the force. There is in fact no need for the initial current that we first postulated. If the circuit loop is made without a battery (Figure 67), movement of the wire still produces a current. Raising the wire as shown gives a current directed anti-clockwise around the loop and a downwards force in the region between the magnet poles. The direction of the current is always such that the force it produces through its interaction with the magnetic field *is in opposition to* the movement of the wire. Currents of this type, which are caused by the relative motion of a conductor and a region of magnetic field, are called *induced currents*.

induced currents

52

As currents do not flow in a conductor without a voltage difference, we can think of the movement of a conductor as producing a redistribution of charge (Figure 68d) which is released when a current loop is provided. Currents may be induced in many ways, of which a few are shown in Figure 68(d—h). While there are many arrangements, just one rule is needed to describe the current flow. Imagine a map of the magnetic field with the current-carrying loop placed with it. In a given position the loop will be linked with a number of magnetic field lines. All of the movements which change the linkage cause a current to flow. No current is induced by a movement which leaves the linkage unchanged. The generality of this rule is confirmed by the fact that with an electromagnet, current may be made to flow by changing the field strength while keeping the conducting loop stationary (Figure 69). For this case it is not even necessary for the loop to lie in a region of large field—it need only enclose it.

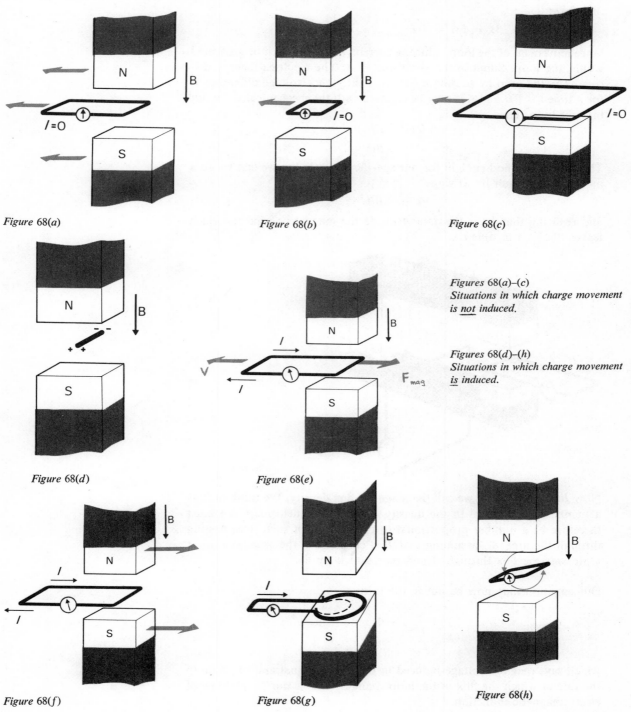

Figure 68(a)

Figure 68(b)

Figure 68(c)

Figure 68(d)

Figure 68(e)

Figures 68(a)–(c)
Situations in which charge movement is _not_ induced.

Figures 68(d)–(h)
Situations in which charge movement _is_ induced.

Figure 68(f)

Figure 68(g)

Figure 68(h)

We can put the idea of induced currents on a quantitative footing very simply by treating our moving current loops as energy converters turning the mechanical energy used in pushing the loop into electrical energy in the form of a current which flows in the loop. We return to the very simplest of the current generating experiments with a rectangular loop (Figure 68e) moving out of a region of approximately constant flux density B. We suppose that motion takes place with velocity v and causes a current I to be induced in a loop of width l.

To keep the loop in motion, a force must be applied. Let its value be F. The work done by this force in a time t is Fvt, vt being the distance the force moves in time t. Now the force required to keep the loop in motion is just equal, but in the opposite direction to, the magnetic force $F_{magnetic}$ on the loop due to the interaction of the induced current I with the magnetic flux of density B. The mechanical work W done in time t is therefore:

$$W = Fvt = -F_{magnetic} \times vt = -BIlvt$$

Since movement of the loop induces a current, a voltage must be generated within the loop. Suppose that its value is V. The electrical energy dissipated in the loop by passing a current I through a voltage difference V for a time t is VIt and this may be equated with the mechanical work put into the loop. Thus

$$VIt = -BIlvt$$
$$\text{or} \quad V = -Blv$$

This answer is good enough for our specific example but we can get to a more general result by writing

$$Vt = -Blvt$$

and realizing that lvt is the area A inside the magnet pole pieces which leaves the loop in time t.

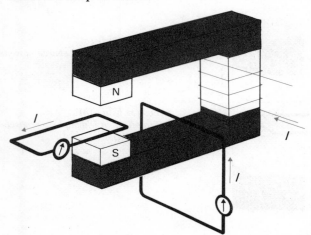

Figure 69

Now B is the quantity we call the magnetic flux density. We think of it as a number proportional to the density of lines on a field map. BA must therefore be a number proportional to the number of such lines passing through the area A—assuming that B is constant in the area. We say it represents the flux through A and represent it by Φ_A.

Our equation may now be put in the form

$$V = \frac{-\Phi_A}{t}$$

which says that the voltage induced in the loop is equal and opposite to the rate at which the flux in the loop changes. This is our general law of electromagnetic induction.

The negative sign in the equation is interesting. It refers, of course, to the law of energy conservation; to the idea that you can never get something for nothing. As the figure is drawn, the movement of the loop reduces the flux linked with it, and the induced current flows, in response to the voltage V, in a clockwise direction viewed from above. If we remember the right-hand screw rule this means that the field caused by the induced current is downwards and in the direction of the field from the magnet. The current is always directed so as to maintain the flux linkage with the circuit loop at its previous value. Just imagine what would happen if this were not so: movement would change the flux linkage and produce a current; the current would produce a field which would *increase the previous change in linkage*; and this would *increase the previous current*; and so on.

Our first task in this section was to find a means of generating electrical currents from mechanical motion and hence of converting mechanical energy to electrical. We found that this could be done by moving a conductor across a magnetic field. Movement of the conductor induced a voltage; this drove a current, and this in turn reacted with the field to produce a force opposing the motion. Mechanical energy was absorbed by movement against the opposing force and electrical energy was produced as a circulating current.

The idea of force obtained from a current and magnetic field was familiar from the previous section; the feature new to the present section is that of currents induced by changing linkages of magnetic field. For the remainder of this section, I want to concentrate on this phenomenon of induction, showing how it leads naturally to the use of alternating currents for the distribution of electric power and underlies much of the electrical machinery we use today.

6.2 Electrical generators

Our first problem is that of finding how to generate electrical power. The devices sketched earlier all produce currents from mechanical motion, but none is really suitable as described. In terms of our earlier discussion, however, we can provide a formula for power generation. Thus is: find a situation in which mechanical means can be used to change the magnetic flux linked with a circuit, then a continuous flux change will give a continuous current.

Quite obviously flux cannot be changed at a steady rate for an indefinite time, so continuous currents may not be developed without some form of switch. We have met such a switch in the commutator of the d.c. motor, and Figure 70(a) and (b), shows schematically how this motor may be used

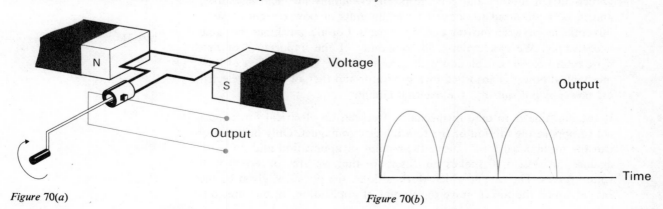

Figure 70(a)

Figure 70(b)

as a generator, an arrangement which was used universally in motor cars before the introduction of the alternator. With a proper d.c. generator, as with the motor, multiple coils and an iron core are used to smooth and increase the output.

For the generation of alternating currents two distinct arrangements of output and field system are possible. Either a coil is rotated inside a fixed magnet or a magnet is rotated inside a fixed coil. Figure 71 shows schematically how current may be drawn from a rotating coil, the field being provided either by a permanent magnet or a d.c. electromagnet. The system used for machines of the type described in the case study on the Electricity Supply Industry is shown schematically in Figure 72. For these large machines direct current is fed to the rotor so that it becomes a spinning electromagnet whose flux linkage with the fixed stator winding reverses at each half turn and develops in them a large alternating current which is the output. In such machines three separate sets of windings spaced

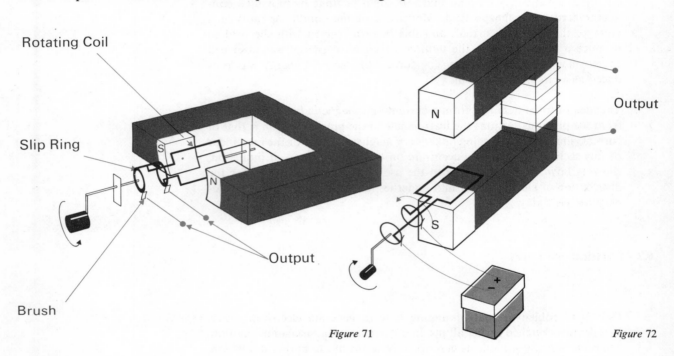

Rotating Coil

Slip Ring

Brush

Output

Output

Figure 71

Figure 72

around the stator develop three similar outputs displaced from each other in time as in Figure 73. An output of this type is what we call a three-phase supply. Permanent magnets do not at present provide sufficient flux to be used for the rotors of these machines, but there is some chance that the strong rare-earth magnets now being developed may be used in this way.

Direct-current motors and generators are essentially identical machines, and in both the need to convert large amounts of power requires large flux linkages between moving and static parts. For a.c. machines the same needs apply. We can understand something of the importance of flux if we return to our calculation with a current loop. We think first of how mechanical power is absorbed by a generator and then of how the electrical power is put out into the external circuits.

If the machine is to take in mechanical power, an electrical force must act to oppose the motion of the rotating electromagnet. Only in this way can the mechanical forces move their point of application and do work against the system. It makes no difference that we are concerned with rotary forces. The rate at which work is done, the power, is given by the rate at which the forces move their point of application. In our previous

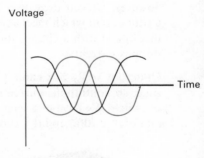

Voltage

Time

Figure 73

linear example the work W was done in a time t, and this corresponds to a power $P = W/t = Fv$. For the machine to absorb a lot of power therefore it must either be turned rapidly or produce a large opposing magnetic force. Since the opposing force is given by F magnetic $= BIl$ we can increase the power by increasing the flux density.

Large flux densities therefore give large opposing forces and permit the machine to absorb mechanical energy at a high rate.

The output of electrical energy by the generator correspondingly requires large flux densities. The flux linkage of the rotating electromagnet with each output circuit is reversed at each half turn. If the maximum linkage is Φ and the rotor makes n revolutions per second, the mean voltage induced in the output circuit during reversal is given by

$$V = \frac{\text{change in linkage}}{\text{time taken}} = \frac{2\Phi}{1/(2n)} = 4n\Phi$$

This is the voltage available to drive currents in the output circuit. With an external resistance of R the electrical power output will be $P = V^2/R = 16n^2\Phi^2/R$. For a generator of given dimensions this clearly implies that an increase of B, and hence of Φ, increases the possible electrical output.

Generators of the type we have been considering, are commonly used nowadays to convert power at rates of about 500 MW. That is, one generator produces a voltage V and a current I whose product (averaged over complete cycles) is the power P, where:

$$P = IV = 5 \times 10^8 \text{ W}$$

When such generators are designed, the optimum values for current and voltage (see the case study on the Electricity Supply Industry) are chosen with reference to the weight and cost of conductors, the flux which can be developed in the iron, and the ease of insulation of the various parts. In a similar way the optimum voltages for the transmission of power around the country are determined by power losses in the wires and the cost of conductors. The voltages which optimize these calculations are in the range 11 000–400 000 V, and evidently represent no optimum for domestic use. What is needed is a device which will accept electrical power as a current at a particular voltage and transmit it as a different current at a different voltage. The device which does this for alternating currents is called a transformer.

6.3 Transformers

The idea underlying the transformer is really very like that used for the generator. A circuit which is linked with an alternating magnetic flux will develop an alternating current: if the alternating flux is developed by mechanical means we have a generator, if it is developed by an alternating current in another circuit we have a transformer.

A very basic type of transformer is shown in Figure 74. An alternating current fed from a generator produces an alternating magnetic field along the axis of a coil. When a second coil is placed near the first, and in such a way that it is *linked by its magnetic flux*, a voltage is induced in it which alternates at the same frequency as the generator.

In a working transformer, the arrangement of the *flux linkage* is less casual than I may have suggested. We can certainly get a large voltage very easily. The voltage induced in each turn of the second coil is added to give the total coil output so we need merely have lots of turns. But as soon as we try to draw current from this circuit the voltage drops. The reason, as with the motor and the generator, is that we cannot transfer power without flux. If large amounts of power are to be passed between the two circuits then the

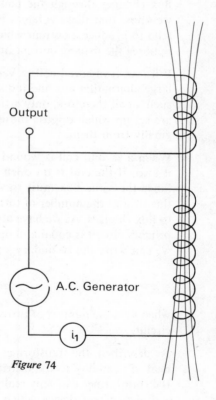

Figure 74

flux linkage

current in the input or primary coil must link as much flux as is possible with the output or secondary coil. Our considerations on magnetic fields show just how this should be done. We merely take an electromagnet for the first coil and its core, put the second coil around the same core and fill up the gap between the polepieces to maximize the flux. With such an arrangement all the flux linked with the primary coil is also linked with the secondary (Figure 75).

The detailed use of flux in a power transformer is interesting. Consider first what happens when a single coil is wound around an iron core and driven by a generator producing an alternating voltage of magnitude

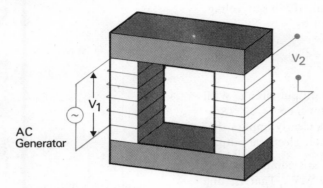

Figure 75

V_1. An alternating current must flow through the coil in response to the voltage, and this current will produce an alternating flux in the core. But the windings of the coil are threaded by this flux and we know that flux changes through the coil must induce voltages in it. Furthermore, we know that these voltages must oppose the driving voltage, since they tend to produce a current which opposes the change in flux and therefore opposes the driving current and its voltage.

With an iron-cored coil, a very small alternating current can produce a large alternating flux and induce a large voltage. When a generator drives such a coil therefore, only a small current flows because induced voltages are set up which oppose those from the generator and differ only marginally from them.

When a second coil is wound on the same core, voltages are induced in it also. If the coil is on open circuit the values can easily be calculated. Since the same flux must traverse the two coils, the *flux linkages* are in the ratio of the number of turns, and so also are the induced voltages due to flux changes. As we have already shown that the induced voltage in the primary circuit is equal and opposite to the input voltage, this means that we can write the secondary voltage as:

$$V_2 = \frac{n_2}{n_1} V_1$$

when n_1 is the number of turns in the primary and n_2 that in the secondary circuit.

As described, the transformer is a very simple device but it gives a great deal of flexibility to the distribution and use of electricity. The merits of the transformer can only really be appreciated after an attempt to design a direct-current device which would do the same job.

6.4 Alternating supplies

The use of alternating currents for the generation and distribution of electricity is clearly attractive, but we need also to reassure ourselves that they are of some use for domestic purposes.

Our major use of electrical power is for heat and light. For each of these purposes the passage of the current through a resistance dissipates electrical power and creates heat. The power being used at any instant is $P = IV$. Reversal of the voltage leaves this unchanged since the direction of the current must change also, making the power on reversal $(-V) \times (-I) = IV = P$. More simply we can remember that $P = V^2/R$ and so has the same value whether V is positive or negative.

Alternating currents therefore produce heating effects comparable with those from direct current, the only difference being that the wire may cool during the short periods of low current. British domestic supplies alternate 50 times per second and so produce bursts of power at intervals of 1/100 second; cooling effects, therefore, are slight. An electric fire cools too slowly for the effect to be discernible at all, but the fine wires of a lamp do cool sufficiently to make the light level fluctuate. With normal supplies you cannot see the effect as a flicker, but on some special low-frequency supplies, such as those on Italian railways which operate at $16\frac{2}{3}$ Hz, the flicker is quite evident. With 50 Hz supplies the fluctuation may be detected by rapid motion. If you move your hand rapidly under an electric lamp its motion can appear to be frozen at the points where the light was at maximum brilliance. A more complete dimming occurs in discharge lamps, and the effect can be dangerous in factories so special measures must be taken to combat it. The trouble arises because rotating machinery often moves in synchronism with the power supply and can have the appearance of being still even when it is in rapid motion.

Electric motors might be expected to present some difficulty in operation from alternating supplies because they are dependent on the direction of current flow. Actually there is no problem except for motors using permanent magnets. The driving force in a motor depends on the product of a magnetic flux B with the currents in a rotor. If the motor uses an electro-magnet driven from the same power service as the rotor, B will be proportional to I, and the driving force, which is proportional to $B \times I$, will be proportional to I^2. The driving force therefore is independent of the direction of the current. The simple direct current motors of Section 5 will therefore work with alternating currents. A number of other kinds of motor have been developed which are dependent on alternating currents for their operation.

These are the main domestic uses of electricity. Since the devices concerned all give satisfactory operation with alternating current supplies we have easily learned to live with them. The appliances which do need direct current supplies are almost exclusively electronic in operation—TV sets and radios—and require little power. For them it is a fairly simple matter, using diodes, to produce a small direct current from the alternating supply.

Self-assessment questions

SAQ 1

What is the resistance of (a) a 40 W electric light bulb and (b) a 1 kW electric fire, both to work off the mains at 240 V? If you measured the resistances of these items when cold, would you expect to obtain these values?

SAQ 2

A power line 10 miles long with an overall resistance of 0·25 Ω per mile supplies a village where the maximum load is 80 kW at 240 V. What will be the voltage at the village when all the load is switched on? How much power will then be dissipated in the line?

SAQ 3

A household contains 20 electric lamps rated at 60 W, 2 electric fires rated at 2 kW each, an electric kettle rated at 1 kW, a washing machine rated at 750 W, a TV set rated at 150 W and an electric cooker rated at 8 kW. What is the maximum current the mains must be able to supply?

SAQ 4

A set of resistors each of 1 Ω are connected together as if along the edges of a cube. What is the total electrical resistance from one corner to the diagonally opposite corner? (Hint—use the symmetries of the arrangement and remember that charge does not accumulate.)

SAQ 5

The potential difference between the terminals of a battery is found to be 2 V when no current is being drawn, but only 1·5 V when a current of 1 A is taken from it. Explain this effect and draw a diagram to represent your explanation. What do you think will be the potential difference when the current is 2 A ? What assumption do you have to make in order to be able to answer this last question? Using the same assumption, what is the largest current you could draw from the battery?

Hint: Batteries present obstacles to the free flow of charge. Treat your battery as if it contained a resistor in series with the cell.

SAQ 6

A moving coil meter has a resistance of 1Ω and a full scale deflection of 1 mA. How could it be used to measure currents up to 1 A? In another application, if the meter is connected in series with a resistance of 999 Ω how will the reading of the meter be related to the potential difference across the meter plus resistance? What sort of meter has it now become? If this combination of meter with resistance is connected to a battery having a potential difference of 1 V at zero current and an internal resistance of 100 Ω, what will be the reading? What can you deduce about the trustworthiness of such meter readings?

SAQ 7

A coil of n turns and enclosing an area A has its ends connected to form a circuit of resistance R. It lies initially in a plane perpendicular to a magnetic field B and is then withdrawn in a time t to a position of zero field.

What is the total electric charge which flows round the circuit as a result. Does it depend on t? Does it depend on whether or not the magnetic flux through the coil decreases at a uniform rate?

SAQ 8

Individual charged particles in motion are equivalent in many ways to continuous currents: an electron moving at a speed is equivalent to a conventional current in the opposite direction. Show that an electron moving at right angles to a magnetic field B should experience a force $F = Bev$. In what direction will this force act, and what path will the electrons follow?

When many electrons move together in a beam, their motion will itself produce a magnetic field. In what way will this field influence the electron motion?

SAQ 9

A beam of electrons is accelerated to a speed v by the application of potential difference V in an electron gun.

Show that $v = \sqrt{\dfrac{2eV}{m}}$ and use your tables to find the speed of electrons accelerated by a potential difference of 10 000 V. The beam passes between two plates parallel to the direction of motion, separated by a distance d. What is the force on each electron when there is a potential difference V' between the plates? What will happen to the beam?

SAQ 10

A square coil of 0·5 m edge with 100 turns of wire is rotated at a speed of 10 revolutions per second about an axis perpendicular to a magnetic field of 0·1 T. Draw a graph to show how the potential difference induced in the coil will vary with time and find an *approximate* value to set the scale of voltage.

SAQ 11

The village mentioned in SAQ 2 is to be supplied by a high voltage cable operating at 32 kV. What should be the turns ratio of the step down transformer in the village? What will be the current in the high voltage cable at full load and the voltage drop in the line? By how much will the 240 V supply drop at full load?

SAQ 12

A highly ionized gas is made to flow at high speed v through magnetic field, B, at right angles to the direction of flow. What will happen to the electrons and to the positive ions? If two metal plates are placed parallel to the flow and to the magnetic field, and on either side of the gas flow, show that the maximum potential difference which can be generated across these plates will be given by $V = dvB$ where d is the distance between the plates. Suggest a possible application of this effect.

Self-assessment answers

SAQ 1

(a) Power used in the bulb is:

$$P = IV = \frac{V^2}{R} = 40 \text{ W}$$

$$R = \frac{V^2}{P} = \frac{240^2}{40} = 1440 \ \Omega$$

(b) For the fire:

$$R = \frac{240^2}{1000} = 57 \cdot 6 \ \Omega$$

Both fires and light bulbs heat considerably in use. These values will be much greater than the cold resistance.

SAQ 2

Effective circuit is shown in Figure 76. The power line provides a resistance, r, in series with the effective resistance, R, of the village.

$$r = 10 \times 0 \cdot 25 = 2 \cdot 5 \ \Omega$$

$$R = \frac{240^2}{8 \ 10^4} = 0 \cdot 72 \ \Omega$$

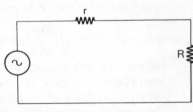

Figure 76

When supplied from the distant 240 V source, the voltage at the village is just $V_{\text{village}} = \frac{R}{R+r} \cdot 240 = \frac{0 \cdot 72}{3 \cdot 22} \times 240 = 53 \cdot 7 \text{ V}$. The voltage drop in the line is $V_{\text{line}} = 240 - V_{\text{village}} = 240 - 54 = 186 \text{ V}$. Power dissipated in the line is $P_{\text{line}} = \frac{V_{\text{line}}^2}{r} = \frac{186^2}{2 \cdot 5} \sim 13 \cdot 8 \text{ kW}$. This is why villages are *not* supplied by 10 mile long lines at 240 volts.

SAQ 3

The maximum power which can be used by the household is:

$P = 20 \times 60 + 2 \times 2\ 000 + 1 \times 1\ 000 + 1 \times 750 + 1 \times 150 + 1 \times 8\ 000 \text{ W}$
$\quad = 15 \cdot 1 \text{ kW}.$

The voltage of the supply can be taken as 240 V, so the maximum current required is:

$$I = \frac{P}{V} = \frac{14 \cdot 95 \ 10^3}{240} = 62 \cdot 9 \text{ A}$$

SAQ 4

We suppose that a voltage V supplied between opposite corners of the cube causes a total current I to flow. Symmetry requires that the ingoing current will split evenly among the three branches at the input point (Figure 77); similarly, the current arriving at the output point is made up from the currents from three similar branches. To the six resistors in contact with an input point or an output point, we can therefore assign a current $I/3$.

At the point marked A, one current is now known to be $I/3$. But the other currents which feed this point are connected symmetrically through the circuit to the input point, and must be equal. To avoid charge build-up,

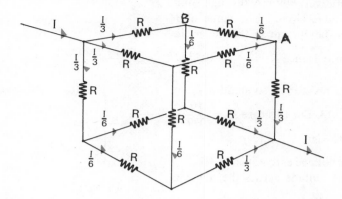

Figure 77

their sum must equal the $I/3$ of the third current from the point, so they are each $I/6$. Applying this reasoning to the other circuit elements shows that all resistors not in contact with an input or output point carry a current $I/6$, while all others carry $I/3$.

To find the resistance of the combination we just find the voltage between input and output points. We do this by adding the voltage along any path joining these points. Beginning at the output, the voltage to point A is $I/3 \times R$, while that from A to B is $I/6 \times R$, and from B to the input, it is $I/3 \times R$. The total voltage difference is thus:

$$V = I/3 \times R + I/6 \times R + \frac{I}{3} \times R = \frac{5}{6} IR.$$

But the resistance of the combination is

$$R' = \frac{V}{I} = \frac{5}{6} R$$

With individual resistors of value 1 Ω, $R' = 5/6$ Ω.

SAQ 5

Charge moves through an electrolyte with the same drifting and colliding motion we met in the movement of electrons through a metal: electrolyte therefore presents an electrical resistance to current flow. We can allow for this internal resistance in a cell by representing our battery as if it were a perfect cell producing 2 V with a resistor r in series as in Figure 78. The battery is accessible electrically only through its terminals which lie outside the cell plus resistor combination.

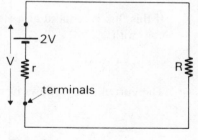

Figure 78

When a current of 1 A passes, a voltage of $1 \times r$ is developed across the internal resistor. This reduces the voltage across the terminals to $2 - 1 \times r = 1.5$ V, as given. Thus r is effectively 0.5 Ω when a current of 1 A passes. If we assume that this resistance is a constant for the battery, we can calculate the voltage between the terminals—called the potential difference or p.d. of the cell—for any current. For $I = 2$ A we expect $V = 2 - 2 \times 0.5 = 1$ V.

The highest current which can pass is that for an external resistance of zero. In this case the cell produces a voltage V in a circuit of total resistance r. The current which passes is $I = \frac{V}{r} = \frac{2}{0.5} = 4$ A.

In the real cell the internal resistance is likely to vary with current, but over a limited current range a calculation like that above may apply.

SAQ 6

The meter will always show full scale deflection for a current of 1 mA. To use it to record a current of 1 A we must divert 999 mA through some alternative path. Figure 79(a) shows a suitable circuit. The value of R must be chosen so that it passes 999 times the current in the meter: its value is therefore $\frac{1}{999}$ times the meter resistance. Hence $R = \frac{1}{999} \sim 10^{-3}$ Ω.

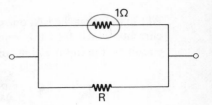

Figure 79(a)

Because it is difficult to make and adjust a resistor of so small a value, we might instead use the circuit of Figure 79(b). For this case $R' = \frac{10}{999} \sim 10^{-2}$ Ω which would be simpler to tailor to size.

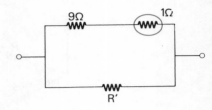

Figure 79(b)

When the meter passes a current of 1 mA the voltage across it is $1 \times 10^{-3} = 1$ mV. With the circuit of Figure 79(c) the voltage across the meter is $\frac{1}{999+1}$ times the input voltage. If 1 V is applied to the combination of meter and resistor, the meter shows full scale deflection: the instrument becomes a 1 V voltmeter.

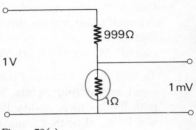

Figure 79(c)

In using this meter to measure the potential difference of the cell, the effective circuit is Figure 79(d). The battery supplies 1 V at zero current, but with the circuit shown there is a current $r = \frac{1}{100+999+1} = \frac{1}{1100}$ A. The meter reads full scale for a current of 1 mA so in this case it will record $\frac{1}{1100} \Big/ \frac{1}{1000} = 0\cdot909$ of full scale and this will be interpreted as a battery voltage of 0·909 V. Moral: a voltmeter, to be accurate, must have a resistance appreciably greater than that of the measured circuit between the same terminals. In a similar way an ammeter must have a resistance appreciably less than the sum of the resistors in its circuit.

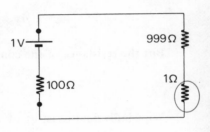

Figure 79 (d)

SAQ 7

The flux initially linked with the circuit is

$$\Phi = n A B.$$

If this flux is reduced at a steady rate a voltage V is induced around the coil, with:

$$V = -\frac{-\Phi}{t} = \frac{naB}{t}$$

The current which flows in response to this voltage is $I = V/R$ so:

$$I = \frac{nAB}{tR}$$

Since the charge Q which flows is related to the current by $Q = I t$ we must have:

$$Q = \frac{NAB}{tR} \times t = \frac{nAB}{R}$$

This answer will apply irrespective of the rate of flux change or of uniformity in this rate. To see why, think of a short time interval Δt during which the flux linkage changes by $\Delta\Phi$. The charge flow in this interval is:

$$\Delta Q = I\Delta t = \frac{V}{R}.\Delta t = -\frac{\Delta\Phi}{\Delta t}\cdot\frac{\Delta t}{R} = -\frac{\Delta\Phi}{R}$$

where I and V are instantaneous values for current and voltage. We can find the total charge flow by adding all the small amounts of charge which move in successive intervals during the flux change.

$$Q = (\Delta Q_1 + \Delta Q_2 + + \ldots) = -\frac{1}{R}(\Delta\Phi_1 + \Delta\Phi_2 + \ldots).$$

The way in which Φ varies with t does not appear in computing the answer and Q therefore is independent of the rate of flux change.

SAQ 8

An electric current is a flow of electric charge. Nothing in this unit has specifically suggested that the idea of current may not be used with very small amounts of charge—even with single electrons. We will calculate the force on an electron in a magnetic field assuming that the path it traverses can be treated as a current-carrying conductor.

Suppose an electron moves with speed v through a distance l in time t. The path is equivalent to a conductor which has passed a current $I = \frac{e}{t}$ in a direction opposite to the electron motion.

If a magnetic field B acts at right angles to the direction of motion, the 'conductor' should experience a force:

$$F = BIl = B\frac{e}{t}vt = Bev$$

and this force will be in a direction perpendicular both to the direction of motion and to the magnetic field in accordance with the left hand rule.

As you may remember from the unit on Mechanics, a free particle which experiences a force perpendicular to its direction of motion moves in a circle. For an electron of mass m, the electromagnetic force we have calculated will produce motion in a circle of radius r, where:

$$F = Bev = \frac{mv^2}{r} \quad \text{or} \quad r = \frac{mv}{Be}.$$

An electron entering a region of uniform magnetic field in a direction perpendicular to the field will move in an arc. Depending on the electron speed, the field strength and the field distribution, the electron may emerge from the field in any direction within the plane of the arc (Figure 80). Conditions which approximate a reflection are associated with the fields used in thermonuclear or plasma research. Smaller angles of deflection, in which the electron penetrates through the field zone, are used in forming the pictures of the domestic television set. This method is preferred to the electrostatic deflection described earlier in the unit because relatively large deflections are more readily obtained.

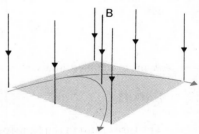

Figure 80

An electron beam, like any other form of current, produces a magnetic field which encircles it in accordance with our right-hand rule for conventional current. Individual electrons moving within this field experience forces directed towards the centre of the beam. We can describe these forces succinctly by noting that electrons in motion are effectively parallel conductors carrying currents: such conductors experience mutual attraction. A beam of electrons subject to these forces would collapse completely, but for the electrostatic forces which hold the electrons apart.

SAQ 9

In an electron gun, electrons are accelerated by an electric field. Since there is a vacuum, they do not make collisions but smoothly increase speed between the electrodes. They move so as to lose potential energy and gain an equal amount of kinetic energy. The kinetic energy is $\frac{1}{2}mv^2$, the potential energy change is eV. Thus:

$$eV = \tfrac{1}{2}mv^2 \text{ or } v = \left(\frac{2eV}{m}\right)^{\frac{1}{2}}$$

From tables $e/m \sim 1\cdot8\ 10^{11}$ coulomb kg^{-1}: a 10 000 V accelerating potential therefore produces an electron speed of

$$v = (2 \times 1\cdot8\ 10^{11} \times 1\cdot0\ 10^4)^{\frac{1}{2}}$$
$$= 6\ 10^7 \text{ m sec}^{-1}$$

Very fast indeed!

Between the plates there is an electric field of magnitude $E = \dfrac{V'}{d}$ directed towards the plate at the lower potential. This produces a force of magnitude $Ee = \dfrac{V'e}{d}$ on each electron which is directed towards the plate at higher potential. With very long plates, the electrons of the beam will accelerate continuously towards the higher-voltage plate and eventually collide with it. With short plates the electrons will again accelerate towards the high-voltage plate and will then move with a constant component of velocity in this direction on leaving the plates: the beam will be deflected.

SAQ 10

Since the magnetic flux threads the coil first in one direction and then in the other, the induced voltage must alternate in direction. Since the change of flux is smooth we can expect the voltage to vary as in Figure 81. To put a scale to the voltage variations we note that in a quarter turn, which takes $\frac{1}{40}$ second, the flux linkage may vary from its maximum value to zero. This is enough to define a mean induced voltage, and will be adequate for the present purpose of setting a scale.

The maximum flux linkage possible is:

$$\Phi_{max} = nAB = 100 \times (0\cdot5 \times 0\cdot5) \times 0\cdot1 = 2\cdot5$$

If this drops to zero in $\frac{1}{40}$ second the mean induced voltage in the interval is:

$$V = \frac{2\cdot5}{1/40} = 100 \text{ V}$$

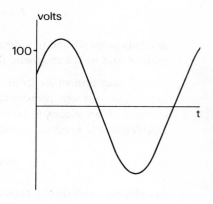

Figure 81

A more detailed analysis would give a maximum voltage of 157.

SAQ 11

The turns ratio in the transformer can be taken as the ratio of the voltages, so the ratio of high-voltage turns to low-voltage turns is

$$\frac{32,000}{240} = 133.$$

Full load is 80 kW. At 32 kV this requires a current:

$$I = \frac{P}{V} = \frac{8\ 10^4}{32\ 10^3} = 2\cdot5 \text{ A}.$$

At this current the voltage drop in the line will be:

$$\Delta V = Ir = 2\cdot5 \times (10 \times 0\cdot25) = 6\cdot25 \text{ V}.$$

The fractional drop in the high voltage line is $\dfrac{6\cdot25}{32\ 10^3}$; this is also the fractional drop in the low voltage line. The actual drop in the 240 V supply is thus $\dfrac{6\cdot25}{32\ 10^3} \times 240 \approx 4\cdot7\ 10^{-2}$ V.

66

SAQ 12

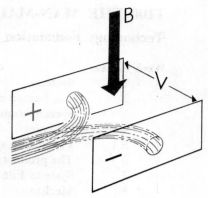

Figure 82

As they move alongside each other the positive and negative ions are effectively equal electric currents moving in opposite directions. On entry to the magnetic field, electromagnetic forces are experienced by individual ions and they tend—as in SAQ 8—to force the ions into circular paths. The forces on the ions must be oppositely directed, by the left-hand rule, so that charge will tend to separate in the beam. Plates placed beside the beam will therefore become charged (Figure 82).

Now ions approaching the charged plates must experience electrostatic repulsion and charge can only accumulate until this repulsion just balances the electromagnetic force.

The greatest electromagnetic force that an ion may experience towards a plate is Bev, the opposing electrostatic force is just $\dfrac{Ve}{d}$ so the maximum potential difference which can be developed is given by:

$$Bev = \frac{V}{d} e$$
$$\text{or } V = dvB.$$

Variations of this arrangement have been suggested for the direct generation of electrical power. The idea is to seed the hot efflux of say a turbo-jet engine with quantities of a readily ionizable material like potassium or sodium. The rapidly moving ions split in the magnetic field, charge the plates and allow a current to be drawn. Although the method in principle seems to be fine, its realization in useful form has not yet been achieved.

T100 THE MAN-MADE WORLD

Technology Foundation Course Units

Week number	Correspondence text		Unit number
1	Systems		1
2	The human component		2
3	Speech, communication and coding		3
4	Statistics	(first part of)	10
5 ⎫ 6 ⎬	Production systems modelling The production environment		⎧ 28 ⎩ 29
7	Systems File		5
8	Mechanics		6
9 ⎫ 10 ⎬	Electricity and magnetism		⎧ 7 ⎩ 8
11 ⎫ 12 ⎬	Energy conversion Power and society		⎧ 20 ⎩ 21
13	Environment File		23
14 ⎫ 15 ⎬	Maintaining the environment		⎧ 26 ⎩ 27
16	Noise abatement		26/27S
17	Economics File		11
18	Structures and microstructures		9
19	Materials		22
20 ⎫ 21 ⎬	Chemical technology		⎧ 24 ⎩ 25
22	Cities File		30
23	Automatic computing		12
24	The heart of computers		13
25	Computer systems		14
26	Economics of traffic congestion		31
27	Transport File		31 File
28	Reliability	(second part of)	10
29 ⎫ 30 ⎬	Analogue computing		15
31	Control		16
32	Modelling II		18
33 ⎫ 34 ⎬	Design		⎧ 32 ⎨ 33 ⎩ 34